CLIMATE IN HUMAN PERSPECTIVE

HELMUT E. LANDSBERG
1906-1985

Climate in Human Perspective

A tribute to Helmut E. Landsberg

Edited by

F. BAER
N. L. CANFIELD

Department of Meteorology,
University of Maryland, College Park, U.S.A.

and

J. M. MITCHELL

National Oceanic and Atmospheric Administration (retired),
McLean, Virginia, U.S.A.

SPRINGER SCIENCE+BUSINESS MEDIA, B.V.

Library of Congress Cataloging-in-Publication Data

Climate in human perspective : a tribute to Helmut E. Landsberg / edited by F. Baer, N.L. Canfield, J.M. Mitchell.
p. cm.
Includes index.
ISBN 978-94-010-5466-9 ISBN 978-94-011-3320-3 (eBook)
DOI 10.1007/978-94-011-3320-3
1. Climatology. 2. Landsberg, Helmut Erich, 1906- .
3. Meteorologists--United States--Biography. I. Landsberg, Helmut Erich, 1906- . II. Baer, Ferdinand. III. Canfield, N. L. (Norman L.) IV. Mitchell, J. Murray (John Murray), 1928- .
QC981.C63 1991
551.5'092--dc20
[B] 90-24124

ISBN 978-94-010-5466-9

Printed on acid-free paper

TABLE OF CONTENTS

PREFACE

The editors intend that this book conveys the remarkable variety and fundamental importance of the late Helmut E. Landsberg's many contributions to the science of climatology and its practice over a very productive 55-year career. We thank the distinguished authors for their contributions. We also thank Corinne Preston and Charlene Mann for their invaluable word-processing assistance and preparation of camera-ready copy. Finally, we thank Joshua Holland for permission to reproduce his portrait of Landsberg, and Jeanne Moody for preparation of the index.

F. Baer
N. L. Canfield
J. M. Mitchell
Editors

CONTRIBUTORS

Ferdinand Baer, Department of Meteorology, University of Maryland, College Park, Maryland, USA

Norman L. Canfield, Department of Meteorology, University of Maryland, College Park, Maryland, USA

Dennis M. Driscoll, Department of Meteorology, Texas A & M University, College Station, Texas, USA

William H. Haggard, Climatological Consulting Corporation, Asheville, North Carolina, USA

David M. Ludlum, Founding Editor, *Weatherwise*, Princeton, New Jersey, USA

Thomas F. Malone, St. Joseph College, West Hartford, Connecticut, USA

J. Murray Mitchell, National Oceanic and Atmospheric Administration (retired), McLean, Virginia, USA

Timothy R. Oke, Department of Geography, University of British Columbia, Vancouver, British Columbia, Canada

Joseph Smagorinsky, National Oceanic and Atmospheric Administration (retired), Princeton, New Jersey, USA

Hessam Taba, World Meteorological Organization (retired), Geneva, Switzerland

Morley Thomas, Atmospheric Environment Service (retired), Downsview, Ontario, Canada.

OVERVIEW

Ferdinand Baer

Helmut E. Landsberg would have been 80 years old on 9 February 1986, and to honor him on that occasion, the Department of Meteorology of the University of Maryland College Park had planned a scientific colloquium on climate to commemorate and highlight his achievements. Unfortunately, Professor Landsberg died on 6 December 1985 while participating in a meeting of the World Meteorological Organization Commission for Climatology in Geneva, Switzerland. The planned symposium was consequently rescheduled as a memorial symposium, and the contents of this volume contain the scientific presentations of that event.

The symposium was, however, more than simply a presentation of Landsberg's scientific accomplishments, although as the reader will discover, those accomplishments were indeed heroic. Landsberg had invested some 55 years in a phenomenally productive career, and in that time he used many opportunities to direct and have impact on numerous individuals and programs. His personal interactions with students and colleagues were succinctly summarized by Werner A. Baum, Dean, College of Arts and Sciences and Professor of Meteorology, Florida State University, who presented the following review at the symposium under the title "*Helmut E. Landsberg: Mentor and Colleague.*"

> Others will speak of Helmut Landsberg, scholar and scientist. I will speak of him as mentor and colleague, as friend and teacher outside the classroom.
>
> Though Helmut and I were both at the University of Chicago in the early 1940's, our paths did not intersect until 1947-48, when we were both spending many hours in a place we both revered, the old Weather Bureau library. For him, I think, it was a haven from his Pentagon office. I was doing bibliographic work as a Research Associate at this university, the University of Maryland, under a U. S. Army contract. We became thoroughly acquainted from hours of conversation. He was most generous with his time and advice, both professional and personal, despite the fact that my major professor was another European climatologist, Erwin Biel, and the relations between the two were somewhat strained.
>
> From those dialogues, I gained an enormous respect for Helmut as a human being as well as a scientist. Apparently there was a mutuality, as there were many consequences of those hours.
>
> In the autumn of 1948, Florida State University decided to initiate activity in meteorology. Not knowing whom to recruit to start the program, the dean wrote for advice from two prominent individuals, Horace Byers and Helmut Landsberg. Helmut called me to inquire as to my interest, and I responded positively, having discovered that the University of Maryland (then under the presidency of a former football coach) was not yet ready for meteorology. Horace and Helmut both submitted lists of prospects to Florida State University, with my name at the head of each list. I have always had the suspicion that both lists were in alphabetical order, but I've never had the courage to check out that suspicion. At any rate, that led to my starting the Department of Meteorology at FSU at

F. Baer et al. (eds.), Climate in Human Perpsective, 1–9.

the wet-behind-the-ears age of 26. I am the father of that quite successful enterprise; history should note that Helmut was one of the two grandfathers.

After my move to Florida in 1949, our collegial associations became too numerous to mention in detail. What is striking to me is how much I learned from those associations and how gracious Helmut was in the gradual transition from the leadership of one generation to that of another, a transition most of us do not make with grace.

We completed a generational cycle. In the 1950's, when the American Geophysical Union began its Russian translation program, Helmut was asked to chair the committee overseeing that activity. He insisted that I be appointed a member, because of my experience with the Russian meteorological literature in the Navy. In the 1980's, when the Board on Atmospheric Sciences and Climate of the National Research Council decided to do a study of problems in the climate data area, I was asked to chair the panel. I insisted that Helmut be appointed a member, simply because no one in the nation knew more about the evolution of the data system than did he. Our report, published in late 1986, was dedicated to Helmut.

Of nonscientific lessons learned from this great "role model," as the current vernacular goes, I want to mention only two.

One lesson was to act with conviction on the basis of your beliefs. Helmut believed in climatology as an intellectual challenge and as a practically useful endeavor, and acted in numerous ways on that belief, when climatology in the United States was at the bottom of the ladder in professional respect among atmospheric scientists. When he was in charge of the climatological division of the Weather Bureau, the prevailing attitude was to send him the castoffs; after all, if an individual was not much good in other ways, he could always add a bunch of numbers and divide by 30 or 31, as appropriate. Fortunately, Helmut lived long enough to see a drastic change. After Sahel droughts, some severe winters in the eastern United States, concerns about carbon dioxide, and so on, there is money and respectability in climatology.

The second lesson concerns the disciplinary life of the scholar turned administrator. Helmut spent at least two decades in demanding administrative posts without ever losing his interest or terminating his activities in his first love; indeed, that is why we had those sessions in the Weather Bureau library. The typical behavior of administrators, especially in academia I regret to say, is otherwise. Admittedly it is not possible simultaneously to be a good university dean or president and carry on a major research effort at the frontier of science; but there is plenty of room between that and losing touch with one's field almost entirely. Helmut set a fine example on this point, one of which I tried to take note in my personal conduct.

My generation of World War II meteorologists, now nearing three score and ten in age, has started the process of fading from the scene. As we do so and reflect on our careers and the good fortune we had to participate in an era of exciting developments in our field, we realize the great debt we owe to those who just preceded us. Largely immigrants, small in number but large in attainment, they brought American meteorology to maturity and they brought us to the enterprise. We are grateful to all of them, and many of us to Helmut Landsberg in particular. Thank you, Helmut!

Landsberg's accomplishments and influence were extraordinary, and they evoked the expected response for citations at the close of the symposium. Indeed no less than twelve agencies sent representatives to honor Helmut E. Landsberg at the symposium and to present citations. Details of these citations and the corresponding agencies has been documented in a report to the *Bulletin of the American Meteorological Society* by Baer

(1986) and will not be repeated herein. However, the moderator of these citation presentations, Dr. Robert M. White, President, National Academy of Engineering, gave a stirring discourse on Landsberg, his impact on people and science, and we repeat here his words.

> We have gathered here to honor and remember one of the true giants of the science and practice of meteorology. Helmut came to the United States at a young age. He was only 28 years old. His career intersected and influenced generations of meteorologists. He certainly influenced mine. His career was a mirror of modern meteorological and climatological development.
>
> I first became acquainted with Helmut in 1950; I had just received my Doctorate from MIT. Helmut had just assumed the post of Director of the Geophysics Research Directorate of the Air Force Cambridge Research Center. He was my first boss. I was a young scientist awed by Helmut. It was an experience I will treasure.
>
> In a reversal of roles several years later I became Helmut's boss. I came to Washington in 1963 as Chief of the Weather Bureau succeeding Dr. Reichelderfer who had been Chief for about 25 years. Helmut was the Director of the Office of Climatology. I may have been his boss in name, but I always had the feeling that I was working for him. There was hardly a serious problem that came across my desk that I didn't talk over with Helmut. His wisdom provided unusual guidance.
>
> When the Weather Bureau was merged with the Coast and Geodetic Survey to become the Environmental Science Service Administration, he became the first director of the Environmental Data Service. He worked closely with me in trying to organize a new approach to the archiving and use of geophysical, not just meteorological data. And as Permanent Representative to the World Meteorological Organization for many years, I worked directly with Helmut in his capacity as President of the Commission on Climatology. He was the first meteorologist elected to the National Academy of Engineering where we worked together. These were marvelous experiences.
>
> Helmut was probably the most honored meteorologist of our generation. We have many outstanding people in our field, but the diversity of his activities earned him honors in so many different ways. He was the recipient of the foremost international award, the International Meteorological Organization Prize of the World Meteorological Organization. He was also the recipient of the President's National Medal of Science. He has held important posts in many different organizations. He was President of the American Geophysical Union; and Vice President of the American Meteorological Society. He probably edited more books and journals than any individual I know. He was editor of 19 volumes of *Advances in Geophysics*; the *World Survey of Climatology*, the *Journal of Meteorology*, and *Meteorological Monographs*.
>
> His publications are staggering in volume and quality. And many of them are in very different fields. He published over 300 papers in a 55 year period. That is a rate of 5 per year, or one about every other month. He was a prodigious contributor. Long before his time, Helmut was deeply involved in questions of earthquakes and earthquake prediction, not unlike another meteorologist, Alfred Wegener, who was deeply involved in the first concepts of continental drift. He was early in worrying about air pollution, long before we became concerned about the problem. His work in bioclimatology was seminal. I was always impressed by the fact that as early as 1946, that's over 40 years ago, Helmut introduced the concept of climate as a natural resource, a concept that took many years to take hold.

> We will miss him, but his legacy will be with us always. Those of you who were his students will remember him as a fine teacher, those of us who were his colleagues will remember him as a fine scientist and a fine human being. And those of you who did not have the privilege of knowing him will know him through his works and his papers which will live long after him.

Landsberg's career began in a modest way with his study of earthquakes, but as he gradually expanded his interests and activities he gained in stature and prominence to the extent that he was acclaimed the "Foremost Climatologist of the World" in 1978 by the World Meteorological Organization (WMO).

He wove his career in and out of the most significant and provocative topics in the geosciences, integrated a variety of disciplines into a coherent body of knowledge and at the same time provoked and directed his community to contribute and advance their science. He had the gift for developing ideas, for encouraging and stimulating people, and for directing and coordinating institutions. There is little indication prior to Landsberg's collegiate career of the prominence he would achieve, although his early life appeared comfortable and free to explore ideas. However, he became a serious student at the University of Frankfurt where he devoured physics, mathematics and geosciences. He subsequently focused his attention on meteorology at the Geophysical Institute where he was influenced by Stuve, Mügge, Linke and Baur, all noted scientists of the time. Ultimately he completed his dissertation in 1930 under the direction of Beno Gutenberg studying earthquake measuring seismographs and spent a year as a postdoctoral assistant to Linke, during which time he set up a weather station in the Rhineland to study the effects of frosts on vineyards.

Perhaps Landsberg's next assignment, a four year period as chief of the Taunus Observatory, whet his interests in climatology as a comprehensive subdiscipline of the geophysical sciences. It was at this post that he was able to study and think about observations of the atmosphere, about the instruments which make those observations, about the three dimensional nature of the atmosphere (by collecting some upper air measurements); and because of the limiting conditions of the location, he had time to study much of the literature then available. Additionally, he extended his understanding of the effects of radiation on atmospheric forcing and gained insight into microscale processes by measurements of condensation nuclei.

At this point in his life (1934) Landsberg uprooted himself from his traditional background and homeland to accept a professional position at the Pennsylvania State University, there to assist in the development of a yet to be recognized meteorology program. With little but energy and enthusiasm, he designed instruments, begged or borrowed others, and thereby set up a model climatological and geophysical station while concurrently creating and teaching new courses.

In 1941 Landsberg joined Rossby at the University of Chicago to develop the Army Air Corps Cadet Training program with a focus on instrument development and usage. At Chicago he began to investigate the impact of the mesoscale on atmospheric processes by studying low-level wind flows off Lake Michigan and their effect on sand dunes. At this point in his career, his interest in directing scientific activities blossomed. Therefore in 1946 he undertook the role of Executive Director of the Commission on Geophysics and Geography, then a branch of the Federal Research and Development Board. In this capacity Landsberg had his first opportunity to bring together the diverse activities of the universities, the private sector, and government agencies. He was able to learn from each of these units and develop his skill at orchestrating both research and operations in a systematic framework.

Landsberg undertook to further develop and utilize the abilities he gained at the Research and Development Board by taking on the direction of the Geophysics Directorate at the Air Force Cambridge Research Center (AFCRC) in 1951. Here he displayed his broad understanding of the geophysical sciences, as well as his skill at coordinating people to undertake unified projects, and at attracting others to study those problems which he foresaw as crucial to our scientific understanding. He set up a well defined, clearly directed in-house research program which stands to this day as a model, and he recruited top talent from the international scientific pool to assist in his plan.

During this period of organizing and directing research, Landsberg continued to develop his personal ideas about climatology and within a few years, he was again challenged by a new opportunity, now to direct the Climatological Services Division of the U.S. Weather Bureau. Here his planning and organizing skills played a major role in serving the United States' involvement in climate related studies and activities, and also in training the entire international community through the WMO. Landsberg centralized the diverse and spotty climate records then available to the National Weather Records Center in Asheville, North Carolina and thereupon coordinated, quantified and modernized those records by developing and implementing computer oriented storage capabilities. Today, most climatic records are available in a centralized computer based system, and this facility which so enhances our understanding of climate and studies related to it are due in large part to the foresight and energy of Landsberg during this period.

Although Landsberg saw the need for centralizing climate data with an eye for efficient and timely distribution on demand, he also recognized the need for diversified climate services to the entire society, since he had become aware of the broad ramifications of climate in human life and its interactions, from biological effects to long term international planning by nations. To implement such services, Landsberg established the State Climatologists Program in which each of the States created an office headed by a State Climatologist which could provide needed climate information and guidance to its constituents. This program was federally supported for some time through Landsberg's efforts, became very popular, and its significance has subsequently encouraged many States to support it.

Landsberg's success at climate data processing and organization prompted the Environmental Science Services Administration (ESSA) to select him to head its newly formed Environmental Data Services component in 1965. In this capacity, he was responsible for all environmental data to include not only climatological data, but also solar and geophysical data, amongst others. He set about this task with the same enthusiasm and energy which he had applied in his efforts with climate data and achieved comparable success.

In 1976 Landsberg became a Professor Emeritus, engaging only in research and of course acting as advisor to many international committees and councils, wherein he had a prominent impact on the evolution of climate oriented studies and activities and in particular on the "World Climate Program." To this day, the WMO and other agencies carry out projects which Landsberg was instrumental in creating.

The development of Landsberg's ideas and his curiosity about our geophysical environment are probably best understood from his writings, since that was the forum in which he communicated most effectively. The interweaving of his professional activities with his research and the evolution of his ideas was certainly a complex process and it is uncertain whether one or the other ever took absolute control. His first published interests revolved around earthquakes and their measurement through seismography. This interest persisted through his early career, but gradually diminished toward the end of the 1930's when he became more involved in other activities and his interests shifted more toward the atmosphere. His experience at Taunus led him to studies in aviation wherefrom he provided some weather related information for pilots, but this period was short lived and he concurrently spent time studying instruments. Although his interest in instruments was maintained throughout his career, his early studies of the galvanometer, the pendulum and the dosimeter led him more into the questions of data quality, analysis and the general need for observations to understand the geophysical environment. To nurture his curiosity for more general concepts, he undertook specific data analyses during the 1930's, analyzing weather maps, microseisms, sunshine at Taunus, condensation nuclei, aerosols and sleet drops, and some temperature distributions.

Landsberg's focus on climate related problems may have been provoked by the fact that he was a very socially aware individual, and he had a dedicated interest in mankind, in searching out ways to improve life on our planet. This concern can be clearly seen in his work on pollution and was probably stimulated by his early observations of condensation nuclei at Taunus. He did some definitive studies on dust during the 1940's, and gradually related his interests to urban climates by investigating metropolitan pollution, looking for anthropogenic impacts, power plant emissions and associating pollution with weather types. This interest reemerged during his later life in academia, when he had more time to undertake detailed research analyses.

Climate became the towering focus of Landsberg's thinking, research, and effort, and this interest became evident already in 1935 with his study on growing seasons. His enthusiasm for climate oriented studies blossomed after that time and by the later 1930's he began a steady stream of publications which continued throughout his life. Early on he evaluated air mass climatology in central Pennsylvania. He then localized his interest to the study of bed temperatures, and emerged with a consideration of cyclone speeds over North America. As his understanding and knowledge grew, he began to composite his material and published his first and classic text on physical climatology in 1941. This text was followed by summary articles on climate in the *Handbook of Meteorology*, the *Compendium of Meteorology*, periodic articles in various encyclopedia, his authorship of the "City Climate" in the *World Survey of Climatology* (L1981-a) and in *Advances in Geophysics*. The wide variety of topics he addressed included national resources, industrial impacts, the biennial pulse, drought, temperature effects and global fluctuations, all in the context of climate and its basic characteristics. The information he provided by his penetrating insights had an immense impact on the meteorological community, which came to realize that climate was the fundamental topic of the future and was *the* discipline to be investigated. Professor J. Smagorinsky, in his article which is included in this book best describes those insights into Landsberg's influence on climatology.

Climatology encompasses a variety of topics so broad that even Landsberg was not able to include them all in his studies. He did however note the significance of many issues and stimulated others to carry out more detailed investigations. Among those areas into which Landsberg delved deeply was that of regional climates. Landsberg recognized that the various geographical regions of the earth endured influences which made their climates different. Correspondingly, he noted differences in climates of various spatial scales. To emphasize this issue, he identified the climatic properties of numerous regions ranging from continents to tropical forests, countries and cities down to the specifics of streets. He utilized a variety of atmospheric parameters to demonstrate the variability of climate regions, describing clouds, rainfall, temperature, and pressure among others to make his point. Toward the latter part of his career, this interest took on a particular focus, and Landsberg spent much of that period probing urban climates in depth. He seemed to have had a special flair and delight in this topic, we note that he accelerated his investigations therein with much gusto in the late 1960's. In relation to urban habitation, Landsberg discussed climate consequences, in particular with reference to planning, he elaborated on the effective use of observations to understand the urban problem, he emphasized the micrometeorological factors, and even included noise in his considerations. With this base of understanding, he projected the ideal "utopian" city and appropriate city planning, discussed the effects of growth on communities from a climatological perspective, hypothesized on the various effects of land use and noted the impact of shelter in relation to urban climate. He observed that his understanding of regional climates was applicable to urban climates, and he identified the unique characteristics of arid zones. All those investigations and results ultimately led to his seminal study of a specific urban area, the planned community of Columbia, Maryland, which he presented in 1979. In the article by Professor Oke which follows later in this book, the reader will discover in detail how Landsberg affected the urban climate community.

Perhaps Landsberg's greatest love was to delve into the relationships between climatology and biological effects. Although he was not a biologist, he saw clearly the impact of climate on man collectively and on the individual organism—note his early discussions on the climate of the bed. His keen interest in and compassion for people surely drove him to study climate in association with human behavior, human habitats, human adaptability and health. This challenge surfaced early in his career, already in the early 1930's, when he was studying nuclei in the air, the impact of seasons on organisms and the occupational diseases associated with the air in working places. His interest in dust also took on biological implications. He went on to study a wide variety of biological-climatological connections, including drugs, U. V. radiation, heating and cooling, housing, air pollution, disease and mortality. He repeatedly emphasized the aspects of climate as they modified human health, noting the wide range of climatic events which control human behavior and how we adapt and are modified by our environment. This interest, which lasted throughout his life, had a profound impact on the bioclimatological community which he stimulated to broaden and accelerate its research activities, and the details are enumerated in the included article by Professor Driscoll.

Despite his boundless energy, as his career matured Landsberg gradually realized that the problems of his chosen topic were too large to be solved by himself alone, and indeed in his lifetime. He therefore began to summarize his understanding of the discipline of climatology and future projections for it intending thereby to attract more resources both in manpower and money. Landsberg recognized early on what we have now come to hold as a burning issue of the day, the potential for climate change. Beginning already in the mid-1960's, he began a steady outpouring of publications on climate change in an effort to provoke our society into a mode which would alert us to the consequences of our technological advances with regard to our environment, and how subsequent changes could influence our climate and consequently our living conditions. He noted how man and his world problems are related to climate, to the human influences on climate and how man induces climate change (this in 1971 in Leningrad), how pollution affects the weather and climate, how global climate might be expected to change; he made projections on the climate of the future and presented numerous articles on comparable issues. His productivity in this arena did not wane to the last and one can only speculate on how much influence Landsberg's concerted effort over the years has had in bringing us to our current frenzied state of study in climate change. Surely no one is more qualified than Dr. J. M. Mitchell to provide details on this aspect of Landsberg's career, and Mitchell provides us with his insights in one of the included articles which follows.

As Landsberg attempted to involve others into future studies of climatology and to predict how climate might evolve in relation to man's activities, his curiosity also drove him backwards in time to search for parallels in climatic history and to learn from others who might have had some insights before him. He was indeed a true bibliophile, a lover of libraries, and seemed to spend all his time in reading. As his enthusiasm for climate change grew, his historical interests grew apace. He looked at the pleistocene for climatological clues, studied the history of thunderstorms, delved into the roots of modern climatology, interpreted climatic records extracted from historical volcanic records and other early geophysical data, and evaluated meteorological observations from traditional records as far back in time as he could uncover. He unearthed records from St. Petersburg, Breslau, New England, and various places in the Americas. There is no doubt that these excursions aided him in his comprehensive view of climate and stimulated his foresight in the expectations of climatic events and impacts which he anticipated far in advance of others. Some charming myths which surround the history of climatology are elucidated in the sequel by David M. Ludlum.

Although the recognition which Landsberg received and so richly deserved has been documented elsewhere, it is appropriate that we review it here. A modest and unassuming man, Landsberg did not shy from expressing his views and graciously accepted acknowledgment from his peers. And there were many, indicating the esteem in which he was held. For his writing skills and critical evaluations he was selected as an associate editor of the *Journal of Meteorology* (1950-1961), an editor of the *Advances in Geophysics* (1952-1977), editor-in-chief of the *World Survey of Climatology* (1964-1985) and chairman of the publication commission, International Society of Biometeorology (1960-1985). For his contributions to science he was elected to the National Academy of Engineering. He was an honorary member of the New York Academy of Sciences and selected fellow of the following learned scientific societies: Royal Meteorological Society, American Academy of Arts and Sciences, American Association of the Advancement of Science (AAAS), American Geophysical Union (AGU), Meteoritical Society, American Meteorological Society (AMS), and the Washington Academy of Sciences. He was a member of a number of other societies which he served with distinction.

The World Meteorological Organization (WMO), recognizing the valued counsel of Landsberg, elected him as President of the Commission for Special Applications of Meteorology and Climatology (1969-1978), to their Advisory Working Group (1978-1981) and to the Commission on Climatology (1981-1985) and he served that institution vigorously. He held several positions in the National Academy of Engineering (NAE), served on the National Advisory Committee on Oceans and Atmospheres (NACOA) (1975-1977) and was a trustee to the University Corporation for Atmospheric Research (UCAR) (1968-1972). He was a councilor, vice-president and chairman of the awards committee for the AMS, and vice president of Section B of the AAAS. His association with the AGU was particularly close and he served that organization as vice-president and president of the section on meteorology (1953-1959) and as vice president and president of the Union (1966-1970).

From the documentation provided, it is clear that these institutions benefited greatly from Landsberg's interaction with them. Indeed the number and quality of awards which he received are a measure of his contributions. From the AMS he received the Outstanding Achievement in Bioclimatology Award (1964), the Charles Franklin Brooks Award (1972) and the Cleveland Abbe Award (1983). From the German Meteorological Society he received the Wegner Medal (1980). From the AGU he received the William Bowie Medal (1978) and from the WMO he received the IMO prize (1979). In 1983 he was awarded the W.F. Peterson Foundation Gold Medal and in 1985 he was honored with the Solco W. Tromp Award by the Enviroscience Foundation. As the ultimate crown for his lifelong service to science, he was bestowed the National Medal of Science by President Reagan in 1985.

It is no exaggeration to say that this kind and gentle man, in his own systematic way, changed the face of climatology and with it our understanding of the atmospheric sciences, the geophysical sciences, how we view our environment and how we will respond to the challenges of the future as our climate inevitably changes. His ability to visualize a problem, plan a solution, search out references, and relate his conclusions to associated relevant issues should be a model to us now and to future generations of geophysical scientists.

THE FOREMOST CLIMATOLOGIST IN THE WORLD

Morley Thomas

"The foremost climatologist in the world" was the accolade bestowed upon Helmut Landsberg in April 1978 by delegates from 58 countries and ten international organizations at meetings in Geneva. As reported in the October issue of the WMO Bulletin, Landsberg was, at that time, retiring from the office of president of the Commission for Climatology of the World Meteorological Organization (WMO). He had been elected to the office eight years earlier but his contributions to climatology and science in general, both outstanding and prolific, had of course commenced decades earlier. Nor did he fade away as he stepped down from his international office; his research and consultative activities continued resulting in more scientific papers and excellent advice appreciated by people in many, many countries around the world.

Landsberg's contributions to international climatology, and especially to the WMO Commission for Climatology (CCl), were outstanding and extended over a forty year period. Inspection of issues of the WMO Bulletin, which began publication in 1952, reveals several feature articles and conference reports written by him, reference to his scientific and organizational contributions in reports and articles by others, reviews of his books and book reviews written by him, notice of his IMO prize in 1979, a feature Bulletin Interview in 1984 and, sadly, his obituary in 1986. Other WMO publications, the Final Reports of sessions of CCl, for instance, are deliberately impersonally written, but wherever Helmut Landsberg was involved, the published reports carry evidence of his direct and indirect involvement.

Landsberg was not only the "Foremost Climatologist" in WMO activities but he was also very prominent in affairs of the International Society of Biometeorology (ISB), especially at their International Biometeorological Congresses, held every three years. He was usually on the program and regularly provided reports of these Congresses for the WMO Bulletin. He was frequently the ISB representative at CCl sessions, especially after he retired from government service. In 1982 Landsberg was awarded a gold medal by the ISB's William F. Peterson Foundation for his work in human biometeorology.

In addition to his research and papers on all aspects of climate and its applications to society, climate change, the human bioclimate, and the history of meteorology and climatology, Landsberg had a global reputation as the author of such text books as *Physical Climatology* (L1941-f and L1958-c) and *Weather and Health* (L1969-c), *The Urban Climate* (L1981-d) as well as sections on climate in such multi-authored texts as the *Handbook of Meteorology* (L1945-a) and the *Compendium of Meteorology* (L1951-b). Over the past 25 years his efforts as editor-in-chief of the multi-volumed *World Survey of Climatology* have been rewarded by its repeated recognition as the most complete and authoritative work of its kind. Clearly, in the international literature of climatology, the name of Landsberg will long remain.

F. Baer et al. (eds.), Climate in Human Perpsective, 11–19.

Memories

Review of Landsberg's writings and accounts of the meetings, conferences and symposia which he so often dominated with his quiet wisdom and wit will bring back floods of memories and personal perspectives to his many friends and colleagues. Although I remember him at the meetings of the old International Meteorological Organization's technical commissions meetings in Toronto during August 1947, I was to know him only from his reputation and writings prior to 1957. That January, CCl-II met in Washington, where on the social side the foreign delegates watched the Eisenhower Inaugural parade from a window at the Roger Smith Hotel and later witnessed the difficulties presented to Washington auto drivers by a two-inch snowfall. Dr. C. Warren Thornthwaite (USA) was the retiring president of the commission and Landsberg was the U.S. Principal Delegate with a large delegation. His common sense, practical and sound scientific approach to such agenda items as climatic classification, microclimatology, derived climatic values and indices, etc. was most impressive to those of us who were younger and less experienced and were grappling with such sectors of climatology and its applications in our own national services. Following the meetings Landsberg had made arrangements for several of us to visit the National Weather Records Center in Asheville, North Carolina. The procedures and methods we saw in action there in a couple of days were most invaluable to those of us involved in introducing modern methods of climate data handling in our own Services.

At the other end of our common time span in international climatology I vividly remember Landsberg at CCl-VIII in Washington during the spring of 1982. Although retired (at least twice) Landsberg not only attended as a representative of the International Society of Biometeorology (ISB) but he also delivered, as the key note scientific lecture, his now-famous "Climatology: Now and Henceforth" (L1982-b). As I had followed Landsberg as CCl president in 1978, I was in the chair during the plenary sessions of 1982 and I was always aware of his presence at one side of the huge meeting hall where observers from the other international organizations were seated. It is not proper however to label Landsberg as an "observer" because we called on him frequently for his background and scientific knowledge and advice. This was given in his usual urbane and witty manner and, should the chairman then make a poor decision, Landsberg was not above so advising him, but after the meeting. During the latter part of the 1982 meetings Landsberg not only conducted a bus tour of his Columbia, Maryland observation station network to illustrate his urban effort experiment but also invited several of his old friends to a small reception at the Cosmos Club. I am sure that memories of the afternoon on tour and the social evening remain with climatologists in every WMO Region of the globe.

Early WMO Meetings

In reviewing Landsberg's career within the WMO organization one must be impressed by his almost classical progression from "invited expert" status at the 1953 CCl session, to Principal Delegate from the United States in 1957, 1960 and 1965, to president of the

Commission in 1973 and 1978 and finally to his position of ISB representative in 1982 and 1985 when he also served as the unofficial senior scientist and advisor to the Commission. Even before the conversion of the International Meteorological Organization into WMO in 1951, Landsberg had participated in meetings of the original Commission for Climatology in Toronto during August 1947 on the invitation of F.W. Reichelderfer, then Chief of the U.S. Weather Bureau. The Commission had been established by IMO in 1929, an event whose 50th anniversary in 1979 Landsberg took great pleasure in drawing to the attention of his colleagues at a Commission session.

But, back in 1947 when a new post-war Climatological Commission met in unison with the other reinstituted commissions, Landsberg's name was already familiar to most scientists present, even the non-English language delegates, on account of his published research and textbook contributions. Climatology, in those days, was at the beginning of a resurgence since it had gained in stature considerably through its contributions to military logistics in World War II. From a rather static, secondary sector of meteorology it was to leap to increased importance through the use of modern statistics first with punched cards and then computer methods. Landsberg had played a major role in developing the use of climatology during the war and afterwards he was the major scientific and administrative contributor to its progress.

For the next several years, until 1954, Landsberg was employed in a broad scientific administrative position by the U.S. Government but he still found time to publish several climatological papers and articles although his principal energies were not directed to international climatology. As already mentioned, he participated in the First Session of the Commission for Climatology (CCl-I) held in Washington during March 1953 but as an "invited expert" rather than as a U.S. delegate. During this session Dr. C.W. Thornthwaite in his presidential address, called attention to Landsberg's concern and interest in teaching and research in climatology, sectors of the subject Landsberg constantly promoted, especially the teaching of climatology in developing countries. By January 1957 Landsberg was in charge of the U.S. Weather Bureau's Office of Climatology and he led the U.S. delegation to CCl-II, also held in Washington. This was my first CCl session and I was most impressed by Landsberg's leadership. At the start of the session he was selected as Chairman of the Committee responsible for dealing with the research and applied climatology items on the agenda. He also delivered a scientific lecture on "Preparing Climatic Data for the User" (L1958-d), the first of many he gave to WMO bodies, and, at the end of the session two weeks later, was asked to serve on a new Working Group on the Guide to Climatological Practices which he subsequently chaired for the next few years. Although drafts for the chapters were written by various climatologists from several countries and at the WMO Secretariat, the Guide, when published in 1960, bore the "sturdy, common sense" stamp of Helmut Landsberg. It was to remain the bible for the standardization of climatological organization, observations, data processing and statistics, the provisions of data and information, applications, etc. for more than 20 years. CCl next met in December 1960, this time in London, England, and Landsberg again led the American delegation and chaired Working Committee B during the session. R.G. Veryard (UK) was president of the Commission from 1957 until 1960, and at CCl-III C.C. Boughner (Canada) was elected to succeed him. By this time Landsberg, still involved in final editing of chapters of the Guide, did not accept membership on any working group himself but he did ensure that some of his senior Office of Climatology staff such as Harold Harshbarger, Julius Bosen and Herb Thom represented the USA on some important groups.

It was in the 1960s that Landsberg found time to make important contributions to biometeorology. He was a valued participant and contributor at the Third International Biometeorological Congress in France during September 1963 and produced several research papers at this time leading to the publication, in 1969, of his book *Weather and Health* (L1969-c) and to his WMO Technical Note, *The Assessment of Human Bioclimate* (L1972-a). At CCl-IV, held in Stockholm during the summer of 1965, Landsberg represented the ISB as well as being the U.S. Principal Delegate and again chairing working Committee B when agenda items of an essentially theoretical or scientific character were dealt with. I particularly remember his efficiency and effectiveness at those meetings since I had been elected to serve as his vice chairman, a position in which I had little to do beyond admiring his excellent chairing of the meetings. At home, in the United States, by this time, the Office of Climatology was being converted into the Environmental Data Service of the new Environmental Science Services Administration and it is a wonder that Landsberg found much time for international climatology as he became responsible for building an infrastructure to deal with all sorts of additional geophysical and solar data at home.

A Question of Name

By 1966 it is apparent that Landsberg felt he had contributed more than his bit to Washington bureaucracy and he moved to the nearby University of Maryland as a Research Professor. He was an organizer for the Fourth International Biometeorological Congress held that summer in New Jersey and wrote an excellent summary of the meetings for the WMO Bulletin. At CCl-V, held in Geneva in October 1969, Landsberg was an American delegate and an active participant in the session. For the previous four years he had been the Commission's Rapporteur on Human Biometeorology and at the meetings presented his report which was published by WMO in 1972. Freed of his responsibilities of administering the environmental data program in his country, Landsberg was again immersing himself in the science of climatology as befitting a Research Professor and so was particularly active at CCl-V in discussions regarding radiation climatology, climatic fluctuations, fundamental climatology, human bioclimatology and the urban/climate relationships.

It was somewhat ironic then, that just a few years after Landsberg had been freed from his national bureaucratic responsibilities, he was unanimously elected president of the Commission at CCl-V for the ensuing four years. Although the commissions meet, as a rule, only once every four years for two weeks and although the secretarial and other bureaucratic work is performed by WMO personnel at the Secretariat in Geneva, it is the president who must maintain an awareness of what the working groups and rapporteurs are accomplishing, represent CCl and sometimes WMO at meetings of other WMO bodies and international organizations, plan for the future and visit Geneva at least once a year to meet with the Secretariat and to chair meetings of the Commission's Advisory Working Group.

Another ironic situation arose early in Landsberg's presidency when, at the 1971 WMO Congress, he was forced to lobby extensively to avoid a movement to eliminate the Commission as a WMO body. Four years earlier Congress had become concerned with the organization of the technical and scientific work of WMO and had established a Panel of Experts to look into the matter and report in 1971. The Panel suggested to Congress that the scientific and technical activities of the Organization might be divided into two sectors - basic activities and applications activities. Discussion at Congress revealed that many delegates believed that since climatology was but a branch of meteorology, the responsibilities of CCl for such basic activities as observations should go to the Commission for Basic Systems and for research to the Commission for Atmospheric Sciences. Since there already were specific commissions for the major applications activities - commissions for Agricultural Meteorology, Aeronautical Meteorology, Marine Meteorology and Hydrology there would be little need to retain the Commission for Climatology. There was, however, a reluctance expressed by several Europeans to allow the commission to disappear. After much lobbying by Landsberg and others, and, incidentally, Landsberg's declaration to Congress "my whole life I have been a climatologist and am not ashamed of that" it was agreed that although the existing Commission for Climatology should be dissolved, a new commission dealing with "all international aspects of applications for which there is no special commission" should be established. When it was proposed to name it the Commission for Special Applications of Meteorology, Landsberg and other lobbyists went into action again to retain the word Climatology in the name. This was successful and the name of the commission for which Landsberg was president became Commission for Special Applications of Meteorology and Climatology (CoSAMC). Also the terms of reference for the new Commission gave it authority for at least advising "other relevant constituent bodies" (i.e. the other commissions) on observational requirements, climate data processing and climate research even if it had lost prime responsibility for these items. It was probably due to the marked increase in attention given to climate, and to the possibility that mankind's activities were changing it, by the world's media - newspapers, radio, TV, etc. that the name of the commission was to change again and again. No change was made at the 1975 Congress but, in 1979, with the establishment of the World Climate Programme, small changes were made in the terms of reference for CoSAMC and it was renamed the Commission for Climatology and Applications of Meteorology (CCAM). But WMO's planners were not yet finished. At the 1983 Congress the name of the commission was again changed, this time reverting to the original Commission for Climatology (CCl). Back in 1971 Landsberg had been concerned regarding the possible loss of not only the name climatology but also the prospect of decreasing attention to the discipline by WMO and the national meteorological services around the world. By the 1980s though he was more than a little amused with the "about turn" traced by the WMO decision makers! For the purposes of this contribution I have mostly referred to the Commission for Climatology (CCl) over the years since I am sure this is the nomenclature Landsberg always used in his thinking about the Commission.

President of the Commission

To use the parlance of a sports writer, Landsberg was the most "natural" president any WMO technical commission ever had. On the scientific side he was already recognized as the world's leading "scientific" climatologist and after more than two decades of experience in dealing with American bureaucracy, as a senior executive in government, there was little he did not know about running a Commission for Climatology. As a result he did considerable to keep climatology as an integral part of the broad field of meteorology in the face of a rather determined effort by some Members of WMO to reduce the science to an insignificant part of meteorology. And he enjoyed excellent relations with members of the Commission in the various countries and with members of the Secretariat. With Ms. Slavka Jovicic, the Secretariat officer assigned to handle CCl, he established an excellent working arrangement which was to serve all exceedingly well for the eight years of his presidency. Landsberg's activities at the 1971 Congress, as CCl president, where he worked behind the scenes both for the Commission itself and for the word "Climatology" to appear in its title have been noted. Later that summer he was one of the principals at a Symposium on Physical and Dynamic Climatology organized jointly by WMO and the International Association of Meteorology and Atmospheric Physics and held at Leningrad, USSR. At the session dealing with climatic fluctuations and modifications, Landsberg drew on his own data and experience in developing urban areas to speak on the evidence for man-made changes of climate. During the next year, 1972, he was one of a few eminent meteorologists who contributed to the principal document submitted by WMO to the Stockholm Conference on the Human Environment. He also represented WMO at the Sixth Biometeorological Congress in the Netherlands and it was fitting that his WMO Technical Note *The Assessment of Human Bioclimate - a Limited Review of Physical Parameters* was published that year (L1972-a). In September 1973 Landsberg represented the Commission at the IMO/WMO Centenary celebrations in Vienna, Austria and delivered one of the feature lectures - "Developments and Trends in Climatological Research" (L1973-a). The following month he was re-elected president of the Commission at CoSAMC-VI held in Bad Homburg, Federal Republic of Germany.

Prof. Landsberg crossed the Atlantic Ocean frequently during the 1970s. After re-election as president he reported to the WMO Executive Committee in the spring of 1974 and in addition delivered a scientific lecture "Drought, a Recurrent Element of Climate" (L1975-f). He was back in Geneva in the spring of 1975 to report for the Commission to Congress and to see the name of his commission changed again - this time to Commission for Climatology and Applications of Meteorology (CCAM) with some editing to the terms of reference. That summer he again represented WMO at the International Biometeorological Congress but this time the meetings were held at his own University of Maryland. In 1976 he prepared and delivered, on behalf of WMO, "Weather, Climate and Human Settlements" (L1976-f) to the UN Conference on Human Settlements which was held in Vancouver, Canada. With both meetings in North America he was saved at least two transatlantic crossings! During the eight years of his CCl presidency however, there were several such trips each year to attend planning meetings, meetings with the Secretariat and representation at meetings of other WMO bodies and other international organizations.

Much was happening in the world of climatology during the years of Landsberg's presidency in the 1970s. In the 1960s, despite the resurgence after World War II, the very words "climate" and "climatology" had slipped to such low esteem that science administrators attempted, and sometimes succeeded, in removing the words from the titles of the national organizations that dealt with the subject. The words lacked glamour, it was said, and so in the United States, the Office of Climatology became the Environmental Data Service in 1964 and, in Canada, the Climatology Division became the Meteorological Applications Branch in 1971. Climatology was still practiced - the science, the data, the applications, but the word was much too old fashioned for the modern "high tech" world. But the media of the world, aided by modern, almost instant, communications, were to publicize the word climate as it never had been before. By the early 1970s drought, famine and death in the African Sahel, the record cold of 1972 in some northern countries, the drought in the Soviet Union and the El Niño in South America the same year were regular news features. This was followed by a poor monsoon season in India in 1974, cold waves in Brazil in 1975, drought and abnormal heat in western Europe in 1976 and extremely cold winters in eastern North America in 1978 and 1979, all with marked effects on the economy and society in general.

The attention and publicity given to abnormal climate and its resulting global economic dislocations provided an opportunity for a person in Landsberg's position to become an "international expert" in the broadest sense - a scientist who made statements and prophecies and demanded much research money. But Landsberg remained remarkably calm about the whole business of climate change. In his 1973 Vienna speech on climatological research he stressed recognition of the great benefits accruing to the world through cooperation in weather observing and the distribution of data and through the use of mathematics and statistics in dealing with the climatic elements which, as he pointed out are not fixed entities. He urged greater attention to mathematical modelling of climate especially to obtain "predictions without timing" as he so aptly called probability values, return periods, contingency values, etc. Also in 1973, his CoSAMC presidential address centered on applications of meteorology and climatology and the need for more understanding and cooperation between meteorologists/climatologists and those who use, or should be using, climate information. His message to the developing countries of the world was direct and to the point - "intelligent application of meteorological information, some of it yet to be gathered, may prevent many of the costly mistakes that have been made in the developed and industrialized areas of the world".

In his second presidential address, in 1978, Landsberg again stressed applications and urged the development and use of more statistical methods and models. In particular he urged the incorporation of meteorological information into economic models. During the 1970s Landsberg did speak and write on climatic change or climatic fluctuations as he chose to call it. Specifically, he contributed to international journals on the Sahel drought and pointed out that "drought within the present climatic regime can be expected and should be planned for" (L1975-k). He was definitely not a sensationalist and repeatedly came back to the statement that "climate is not a constant factor on either the global or local level". He repeatedly stressed the need for data, for facts and better models in order to explain observed climatic fluctuations since it would be only with this information that it would be possible to predict variations or abnormalities in climate.

Senior Scientist

In April 1978 I succeeded Landsberg as president of the Commission and for the next four years he was a valued member of the our Advisory Working Group. We met twice in Geneva - I remember vividly meeting the Landsbergs one wintry Sunday morning in a transit waiting room at the Frankfurt Airport when even the discomforts of an all night trans-Atlantic flight did not dim the Landsberg wit - when he was the recognized senior advisor to the Commission. He was not one to push his advice onto his international colleagues but, as agenda items were discussed, he could be counted on for both some good historical knowledge and advice for the future. In particular, during the December 1981 meeting of the Group, when it was decided to take advantage of Prof. Landsberg's knowledge and enthusiasm for climatology we prevailed upon him to prepare and deliver a special scientific lecture at the forthcoming Washington meetings. This he agreed to do and the result was "Climatology: Now and Henceforth", subsequently published in the WMO Bulletin (L1982-b).

It was my impression that Landsberg was quite happy to be free of bureaucratic responsibilities during these years. He participated in the World Climate Conference held in Geneva in February 1979 and although not a featured speaker, spoke out from time to time with wit, humour and common sense. Later that year he was honored by WMO with the awarding to him of the twenty-fourth IMO Prize. The presentation was made in Geneva, in November 1979, at the time of a meeting of the Commission's Advisory Working Group and the climatologists present were very proud of this recognition - the first IMO Prize given to a climatologist. At about this time Landsberg was deep into his research into the history of early international meteorological observing networks which was to result in his article - "A Bicentenary of International Meteorological Observations" published in the WMO Bulletin in October 1980 (L1980-a).

A careful reading, and re-reading, of Landsberg's "Climatology: Now and Henceforth" (L1982-b) is a must for every climatologist in every country. His brief account of "mean value climatology" of the past, the deficiencies of current observations and mathematical models and the promise of "satellite meteorology" for indirect sensing of current and new climatic elements was considered, concise and very much to the point. With regard to climatic predictions he stressed again the need to first obtain a better understanding of the causes before the full power of modern computing systems can be most effectively utilized. For, as he stated "ultimately all hypotheses and models have to be tested by observations. These have always been the mainstay of climatology and the core of international cooperation."

The need for a better understanding of the physical theory of climate was again his thesis when he was an invited scientific lecturer at the 1983 WMO Congress. Entitled "The Value and Challenge of Climatic Predictions," his lecture was subsequently published by WMO (L1985-c). In the lecture he admitted that climatic forecasts were still inadequate but he did maintain a hopeful attitude towards predictability. He dealt with the astronomical approaches to climatic variability and described the persistence and teleconnections avenues in statistical models. He stressed the absolute need to incorporate oceanographic data into climate models, not only sea surface temperatures but, in time, knowledge of the deeper ocean circulations. This he considered one of the major challenges for the future. Finally, in 1982, he was honored with the presentation of a gold medal by the International Society of Biometeorology for his many contributions in the fields of human and urban biometeorology and it was when representing the Society at the 1985 session of the Commission for Climatology in Geneva that he collapsed and died.

In retrospect I now realize that Landsberg was always considered to be a "senior scientist" even by those of us who were not much more than a decade younger. He spoke authoritatively and he was authoritative. No one knew the subject of climatology as he did and no one could express himself better than Landsberg in explaining it all. Internationally, the prospect that he was to participate in a meeting or conference automatically increased the attendance. In universities and weather services the world over his name is immediately recognized and his writings cherished. Helmut Landsberg was indeed "The Foremost Climatologist in the World".

HELMUT E. LANDSBERG: TOWARDS AD 2000

Thomas F. Malone

As an eminent climatologist, Helmut Landsberg was recognized worldwide as a scholar in interpreting the past. As a perceptive and creative scientist, educator and administrator, his attention was by no means constrained to retrospection. His fertile mind was continually focussed on the future. Nor were his thoughts on the future restricted to a simply extrapolation of contemporary trends. He had a remarkable capacity to differentiate sharply between a mere projection of what probably would be and a vision of what possibly could be. He was skilled in discerning the steps that should be taken to transform what would be to what could be! Indefatigable energy and remarkable tenacity made him a pre-eminent agent of change in everything he touched. The present status of the atmospheric sciences owes much to this aspect of his life.

Clearly, we cannot fathom what Landsberg would have told us about "2000 AD and Beyond" if he were still with us today. It would be audacious to speculate on what his vision might have been. It would clearly be out of order to substitute our own vision for his. It may be appropriate, however, to identify several characteristics of his scientific and professional life and infer from them the kind of goals we might pursue to honor his memory in an appropriate manner.

First, we might recognize his role in transforming climatology from a somewhat sterile compilation of statistical data into an integral part of a rapidly emerging atmospheric science. Elsewhere in this volume, Dr. Smagorinsky describes with grace, elegance and authority the evolution of meteorology over the past several decades from the artful practice of empiricism to a quantitative science based on well-known physical laws. Climatology could well have been relegated to an inferior position in the atmospheric sciences. That this did not happen is due in no small part to Landsberg's influence. In the early thirties, the European community was considerably more sophisticated in the science it brought to bear on atmospheric issues than was the American community. In his doctoral studies in geophysics and meteorology at the University of Frankfurt, in his early seismic investigations, and in his research on condensation nuclei, thunder clouds, sunshine and aviation meteorology, Landsberg brought a decidedly thorough scientific point of view. This approach carried over to his publication on climate which began before he left the Taunus Observatory to come to the United States. Landsberg was one of that small band of European meteorological scholars who crossed the Atlantic to guide and enrich the burgeoning academic interest in this field in the United States. From 1934 to 1941, he helped lay the groundwork for one of our premier academic departments at Pennsylvania State University. During his tenure at the University of Chicago from 1941 to 1943, he contributed to the intellectual ferment stimulated by Carl-Gustaf Rossby. Then followed a number of years in public service in responsible positions as a meteorological consultant and operations analyst in the military and as Director of Climatology in the U.S. Weather Bureau and then as Director of the Environmental Data Service in the Environmental Science Services Administration. Here

F. Baer et al. (eds.), Climate in Human Perpsective, 21–28.

again, by personal example and leadership, he was an effective force in establishing the scientific and professional tone of governmental service. His insistence on rigorous scientific standards for research was matched by a deep commitment to the application of science to societal needs. The seminal paper he and Woodrow Jacobs prepared for the *Compendium of Meteorology* (L1951-b) drew on their military studies during World War II and set a course for Applied Climatology for years to come. His studies of urban climate, the utilization of climatic data in agriculture, housing and highway planning, and in issues concerned with flood insurance, were landmarks. He took pride in his designation as a Certified Consulting Meteorologist by the American Meteorological Society. Even his somewhat whimsical papers on bed temperatures (L1938-f) and "A Note on Bedroom Bioclimate" (L1962-g) rested on sound scientific investigation. His observations on the role of human activity in changing the climate were a precursor of current activity in that regard.

In sum, Landsberg was one of the prime movers in uniting climatology and meteorology and in transforming them into the atmospheric sciences. When the President of the United States presented him with the National Medal of Science in 1985, a half century of creative scholarship in pure and applied science was appropriately recognized.

We can only stand in awe at the ease and authority with which Landsberg crossed disciplinary boundaries. Paper by paper citation would quickly become tedious. His editorship of *Advances in Geophysics* for a full quarter of a century and his selection as President of the American Geophysical Union in 1968 (Bowie Medalist in 1978) provide ample testimony to his catholic interests and capabilities. My search in 1976 for a wise and knowledgeable chairman of a panel on geophysical predictions, and editor of a monograph in the series Studies in Geophysics, began and ended with one name: Helmut Landsberg.

Nor were his interests and contributions confined to geophysics. Once again, he seemed to have been prescient in anticipating current attention to the possibility of linking the physical sciences with the life sciences. He introduced a course in bioclimatology at Pennsylvania State College in 1940 which was followed by a number of publications in this field. He was elected to the presidency of the American Institute of Medical Climatology, 1969-1980, and received the William F. Petersen Foundation Award for outstanding accomplishments in biometeorology in 1982. In a very real sense, Landsberg's concern personified the perceptive comment of Dr. Lewis Branscomb (1986), chairman of the National Science Board from 1980 to 1984, that science is becoming reunified in that the really exciting work in many fields is being done by people who have demonstrated that they have mastered the ideas and techniques of their original discipline and the capability of bringing the concepts, tools and other disciplines in the resolution of problems that transcend disciplinary boundaries.

Another facet of Landsberg's professional work that is relevant to the direction in which our field is moving was his dedication to the international aspects of the atmospheric sciences and climate. It was only natural that a climatologist would have a global view. His published papers ranged across precipitation fluctuations in the Asian Monsoon over the past hundred years, and urban planning in arid zones and the reconstruction of northern hemisphere temperatures from 1579 to 1880, to temperature observations at Breslau from 1710 to 1721. Active in the World Meteorological Organization, he passed away in 1985 at a meeting of the Commission for Climatology over which he had presided as President from 1969 to 1978. He was an active participant in the General Assemblies of the International Union of Geodesy and Geophysics. His commitment to his responsibilities at the IUGG General Assembly in Moscow in 1971 afforded me the pleasant opportunity to escort Frances Landsberg to the Moscow Circus. Helmut was detained at a committee meeting and I was free, so we enjoyed a memorable evening together.

Landsberg's career in many ways personified the eloquent statement of Lewis Thomas (1984): "I believe that international science is an indisputable good for the world community, something to be encouraged and fostered wherever possible. I know of no other transnational human profession and I include here the arts, music, law, finance, diplomacy, engineering and philosophy, from which human beings can take so much intellectual pleasure and at the same time produce so much of immediate and practical value for the species. The problems that lie ahead for the world, endangering not only the survival of our kind of creature, but threatening the existence of numerous other species are proper problems for science. Not to say that science will find answers in time, simply that there is no other way to solve these problems."

I have identified for special reason Landsberg's role in a) bringing meteorology and climatology into the mainstream of American science, b) opening up the channels of communication among the several disciplines of the earth sciences and between the earth sciences and biology, and c) advancing the international dimension of the atmospheric sciences. Were he with us today to address the issue of "AD 2000 and Beyond" it seems likely to me that he would offer some sage remarks on these three topics, since each is in a significant state of flux at this time and each presents challenges that he would be quick to identify and point out to us. We cannot pretend that the consideration we will give these three topics would necessarily reflect his own views, but reflection on his distinguished career leads us to believe that these topics were dear to his heart and he would want us to address them as thoughtfully as we can in a determined effort to build on the splendid base he has established and move toward the year 2000 with optimism and hope.

With respect to science, it is clear that, as Roland Schmidt, chairman of the National Science Board from 1984 to 1988 has pointed out, we are embarking on a new era of science policy. It is a period in which the "social contract" between the scientific community and government, outlined forty years ago by Vannevar Bush (1945), is now up for review. There are several indications. Congressman Don Fuqua, retiring Chairman of the House of Representatives Committee on Science and Technology, has initiated a major study "American Science and Science Policy Issues." This effort reflects the clear realization of the central role of science and technology in our national life. It is especially timely because it is apparent that there is an urgent need to reverse the growing trend of huge deficits in the federal budget. This wide-ranging effort will include:

- •The goals and objectives of national science policy.
- •The institutional framework of national science policy.
- •Education and manpower.
- •The impact on the information age on science.
- •The role of the social and behavioral sciences.
- •The regulatory environment and the scientific research.

•Funding levels.
•Funding mechanisms.
•The role of the Congress in science policy making.

The purview of this effort is so wide and its implications are so profound that it behooves each one of us as members of the scientific community to become informed and articulate on the litany of issues under each of those topical headings. Such vital issues as the key role of the American university system, in the research enterprise, the serious erosion in federal support for maintaining an up-to-date infrastructure in our colleges and universities, the proper balance between support for individual researchers and research centers, innovations in federal funding of research which bypass traditional review processes, and the role of scientific research and development in maintaining our national competitiveness in an increasingly global economy, are among the topics that are likely to be included in the study. An effort of this magnitude will require several years for completion, but a Chairman's Report was issued in December, 1986. The purpose in pointing out this example of the ferment in science policy is to encourage you—as I know Landsberg would have done—to become informed on the issues and provide an input to this process as it proceeds.

My point in citing this initiative is to draw your attention to it, as I believe Landsberg would have wished, and to emphasize the responsibility of each of us to become informed on these matters and to convey our views to policy makers in both the public and private sectors. In the long run, this will be far more useful in charting a course to the year 2000 than would fanciful speculations on what that year might reveal. During the 1950's and 1960's we were fortunate to have such scientific statesmen as Lloyd Berkner, Alan Waterman, and Paul Klopsteg who perceived the scientific potential and the societal implications of the atmospheric sciences and who were largely responsible for the infusion of something on the order of a billion dollars of NSF funding over the last twenty-five years into our field. It is unlikely that many of us will be able to emulate their contribution to science policy but a broad base of researchers informed and articulate on the science policy issues of this new era is our best insurance that the course selected will be a wise one.

With respect to the role that interdisciplinary research is likely to play over the next decade or so, there are several indications that this issue will loom larger than it has in the past. The study of the atmosphere is inherently interdisciplinary. The most profound influence on integrating climatology and meteorology over the last several decades has been the introduction of quantitative methods for dealing with the dynamics of air motion. A rapidly rising matter is the role of chemistry in atmospheric behavior. But this cannot be resolved satisfactorily without recourse to the interface between living matter and the atmosphere. This necessitates an intimate interaction between atmospheric scientists and ecologists. The need for this interaction was highlighted in an address by Professor Garland, President of the International Union of Geodesy and Geophysics, at the General Assembly of the International Council of Scientific Unions in Cambridge, England, commemorating the twenty-fifth anniversary of IGY. In commenting on the heartening progress that has been made in understanding our planet since the IGY, he went on to remark that many of the remaining mysteries of the earth and its environs lay in the interaction at the interface between the biological and physical world—including man himself. At a similar observance during the Annual Meeting of the National Academy of Sciences, Dr. Herbert Friedman, Chairman on the Commission on Physical Sciences, Mathematics and Resources called for a "bold, holistic interdisciplinary program" to deepen and strengthen our understanding of the total earth system.

Today we stand on the threshold of a renaissance in the sciences concerned with planet Earth and the fauna, flora, and humans that inhabit it. A conceptual framework is emerging in which it is recognized that the terrestrial environment and the diverse forms of life that inhabit it constitute an integrated system of interacting components. No single part—oceans, atmosphere, or biota—can be understood adequately enough to anticipate its changes by treating it in isolation. Similarly, no geographic segment can be analyzed satisfactorily as a sequestered entity.

In the United States, the National Research Council has challenged the scientific community to build upon the rapidly advancing knowledge about our planet and an explosively developing technology to frame a new multidisciplinary paradigm for research in geophysics and global biology. The National Aeronautics and Space Administration (NASA) is initiating a new program on Earth System Science that will involve a deeper understanding of the Earth as an integrated system of interacting components with global changes that can have a profound impact on humanity. The National Science Foundation (NSF) has a bold new emphasis in Global Geosciences, and interest is also evident in other national scientific communities.

The international scientific community—particularly the International Council of Scientific Unions (ICSU), a nongovernmental scientific organization that may be thought of as constituting the world scientific community—is also becoming involved with a new, integrative program exploring natural processes that affect the total Earth system. The groundwork for the program had been laid in 1983, when ICSU declared in an unpublished interim report that "a central intellectual challenge of the next few decades is to deepen and strengthen our understanding of the interactions among the several parts of the geosphere and biosphere." An understanding of the relevance of this endeavor to societal issues—such as anthropogenic impact, biological productivity, and management of global life-support systems—led ICSU to initiate plans for collaboration among scientists and governmental and intergovernmental agencies throughout the world. The objective of ICSU's proposal for the International Geosphere Biosphere Program (IGBP) is:

> To describe and understand the interactive physical, chemical, and biological processes that regulate the total Earth system, the unique environment it provides for life, the changes that are occurring in that system, and the manner by which these changes are influenced by human actions.

The intellectual scope of IGBP extends from the interior of the Earth to the interior of the sun. The hallmark of IGBP is integration—breaking down the barriers that have traditionally compartmentalized the study of the atmosphere, oceans, solid part of the Earth, solar-terrestrial interaction, fauna and flora, and humankind. Both observational and analytical, the nature of the program will require observations of parameters not now being measured and will necessitate an innovative style of interdisciplinary analysis. The most challenging aspect of the proposal is the intent to illuminate the intimate linkage between the physical sciences and the life sciences. The viewpoint is holistic and global.

Both the American Meteorological Society and the American Geophysical Union have established committees on Global Change. The National Research Council (NRC) already has a committee under the chairmanship of Jack Eddy and if the program moves ahead as expected the NRC will set up a special interdisciplinary group to provide input into the ICSU program.

As exciting as this new endeavor may sound, we should not underestimate attendant difficulties. First of all, the infrastructure in our university research community definitely favors individual investigators. The reward and advancement structure is very much discipline oriented. It often seems as though there is a distinct penalty imposed on the young researcher who consciously aspires to an interdisciplinary research effort. The departmental structure clearly favors research in a discipline. Second, the structure of governmental agencies in this country is very much along mission-oriented lines. Although the Office of Science and Technology Policy (OSTP) has made a valiant effort to fashion a structure that would be supportive of this kind of effort, it remains to be seen whether proper orchestration of the existing agency structure can be achieved. This may turn out to be crucial with respect to adequate funding. Third, there is the substantive problem of achieving a harmonious interaction between the physical scientists and the ecologists. Both have distinctly different traditions in the manner in which they go about research, and incompatibility in temporal and spatial scales will present formidable obstacles. These difficulties are not mentioned to dampen enthusiasm for this undertaking—they are presented simply to portray a realistic picture of the problems that will have to be resolved.

In spite of the eloquent words of Lewis Thomas (1984) on the importance of international science—to which we could all subscribe—there are issues that will have to be resolved at the international level if a program of this nature is to success. There is, in my mind, a sort of institutional crisis at the international level. We have been notably successful in forging a working relationship between the nongovernmental International Council of Scientific Unions and the intergovernmental World Meteorological Organization in the case of the Global Atmospheric Research Program (GARP) and the World Climate Research Program (WCRP) in spite of the asymmetry between the modus operandi and funding of an intergovernmental organization and a nongovernmental organization which has inevitably produced stresses and strains. Moreover, the structure of the United Nations Specialized Agencies partakes more of the mission orientation of federal agencies in our country with all the attended problems that we have already noted. Current difficulties in maintaining the support of sovereign nations to such organizations as UNESCO, which in the ideal world might be likened to NSF, do not augur well for employing that mechanism.

A serious effort has been made by the International Council of Scientific Unions to examine its structure and funding and to adapt to the increasing internationalization of the scientific enterprise. The leadership of the Council held a three-day meeting at the Ringberg Conference Center. Nearly two dozen distinguished scientists from outside the framework of ICSU attended as guests of the Council. In a mixture of working groups and plenary of sessions they addressed these questions:

- What should be the future orientation of ICSU to allow it to attain its full potential in the next decade?

- How will ICSU make this orientation a reality?
- How will ICSU insure a process of continuing critical review of its activities and structure?
- How can ICSU prove the effectiveness of communication between its members and its governing bodies?
- How can ICSU mobilize the financial resources and a professional staff needed to carry out its agreed objectives?

The conclusions were decidedly upbeat but much remains to be done by way of institutional nurturing if ICSU is to achieve the role it proposes to fulfill. Its earlier success with the international endeavor known as the International Geophysical Year (IGY) leaves one cautiously optimistic.

It would be quite compatible with Landsberg's accomplishments were he to have charged us with:

- Completing the merging of climatology and meteorology into the atmospheric sciences and taking our place as an honored member of the scientific and technological enterprise. A corollary to this would be to be attentive to the policy issues which will affect our field just as it affects all of science and to raise our voices in an appropriate and timely fashion.
- Breaking new ground and cultivating and embracing the kind of interdisciplinary effort which is required to complete our under-standing of the processes that make planet Earth habitable. A corollary to this is the resolution and wisdom to surmount the institutional and traditional obstacles that stand in the way of such an endeavor.

Tackling the job of putting in place the institutional arrangements at the international level which are vital to the fulfillment of the kind of objectives that Landsberg pursued so assiduously during his distinguished scientific career.

There are no guidelines for the kind of task to which we properly would commit ourselves if we wished to honor the memory of Helmut Landsberg. By way of general advice, I think that Landsberg would have been comfortable with a guidance that I have found useful over the years. It draws on the admonition of the Immanuel Kant set forth two hundred years ago in his book *Critique of Pure Reason*. He said that human reason intends to focus on three fundamental questions:

- First, what can I know?
- Second, what ought I to do?
- Third, what may I hope?

Above all, Helmut Landsberg was an inspiringly hopeful individual. When I think of the courage and optimism with which he faced challenges in his career, I am reminded of the words of Clara Park (1982):

> Hope is not the lucky gift of circumstance or deposition but a virtue like faith and love to be practiced whether or not you find it easy or even natural because it is necessary to our survival as human beings.

References

Branscomb, L. M., 1986: The Unity of Science (editorial). *American Scientist*, **74**(1), 5.

Bush, V., 1945: *Science, the Endless Frontier*. U. S. Government Printing Office, 184 pp.

Park, C. C., 1982: *The Siege; The First Eight Years of an Autistic Child with an Epilogue Fifteen Years Later*. Little Brown, 328 pp.

Thomas, L., 1984: Scientific Frontiers and National Frontiers: A Look Ahead. *Foreign Affairs*, **62**(4), 966-994.

CLIMATOLOGY'S SCIENTIFIC MATURITY

Joseph Smagorinsky

Although I was, in many ways, an outsider to Landsberg's immediate circle of colleagues, our paths crossed many times. The encounters were, in part, the result of our having worked in the same organization. But that will not be the subject of my remarks.

Much more germane was the ongoing evolution in the field of climatology that drew me, and others like me, into the company of the climatological community. This coalescence over the past 30 or 40 years was coincident with the maturing of the field's scientific foundations. It was a transition in which Landsberg played an important spiritual role as a leader and a scientist.

But let me first disclose a few facts about Landsberg which are relevant and were new revelations to me.

The year that Landsberg received his doctorate from the University of Frankfurt, 1930, he already had five papers to his credit. But they were not in climatology. They were the beginning of a long string of publications connected with earthquakes, with only occasional excursions in the following years into meteorological questions. In 1933, he began to digress into incidental climatological discourses.

The cutting edge of climatology at that time we might today describe as descriptive geography. Climatology was essentially a qualitative science, as was meteorology, in general. The quantitative aspects of climatology were the compilations of copious statistical time series of local and regional climatic structure, mainly driven by the needs of agriculture. Climatology was therefore, essentially a continental discipline which stopped at nose level. Even the interests in external happenings: precipitation, cloudiness and sunshine, were primarily for their agricultural impact.

This is the field into which Landsberg entered and made his legendary career impact. It is interesting to note, however, that with only one exception that I know about, his curriculum vitae does not show job titles which carried the word "climatology" or "climatologist". One can find "Assistant Professor of Geophysics", "Associate Professor of Meteorology", "Director, Institute for Fluid Dynamics and Applied Mathematics", "Member, National Academy of Engineering" and "Director, Environmental Data Service". The exception which provided the clue to his true disciplinary identity was when he was Director of the Office of Climatology at the Weather Bureau between 1954 and 1965.

But I think it fair to say that Landsberg, especially in his later years, thought of himself first as a meteorologist, and then, more specifically, as a climatologist. For example, he listed himself as a meteorologist in *Who's Who in the World* (1982). Also, in a 1955 article, he wrote under the title *Climatology - a Useful Branch of Meteorology* (L1955-a). These and other evidences distinguished Landsberg, among other things, in his perception of climatology as a branch of meteorology rather than geography.

On my bookshelf are several volumes on climatology acquired in my early student years, mainly during World War II. They say something about the field at that time. To begin with, there is the third edition, dated 1942, of the 1936 classic by W. G. Kendrew,

F. Baer et al. (eds.), Climate in Human Perpsective, 29–35.

The Climates of the Continents. By its very title, its intention is a descriptive regional climatology. But Kendrew did devote 4 of his 473 pages to the surface manifestations of the general circulation in a chapter titled "Pressure and Wind Systems".

At the beginning of World War II, Landsberg seemed to have developed an interest in some of the more physical aspects of climate and their relationships to the component weather processes and events. Among other things, in 1941, he produced the first edition of his well-known textbook *Physical Climatology*, which was to go through four printings by 1947, and a second edition in 1958. I couldn't find a copy of the first edition, but I'll return to the 1958 edition shortly.

And then there was the landmark of that era, the 1941 article by C. G. Rossby in *Climate and Man* titled "The Scientific Basis of Modern Meteorology". Rossby starts the article with the following:

> The science of meteorology does not yet have a universally accepted, coherent picture of the mechanics of the general circulation of the atmosphere. This is partly because observational data from the upper atmosphere still are very incomplete, but at least as much because our theoretical tools for the analysis of atmospheric motions are inadequate. Meteorology, like physics, is a natural science and may indeed be regarded as a branch of the latter; but whereas in ordinary laboratory physics it is always possible to set up an experiment, vary one factor at a time, and study the consequences, meteorologists have to contend with such variations as nature may offer, and these variations are seldom so clean-cut as to permit the establishment of well- defined relationships between cause and effect ... until proper theoretical tools are available, no adequate progress will be made either with the problem of long-range forecasting or with the interpretation of past climate fluctuations.

In a nutshell, Rossby expressed his dilemma, as seen from his 1941 perspective, and then went on to lay out a strategy for the future. This, to me, is the beginning of the modern era of climatology that I alluded to in my title.

In 1944, Haurwitz and Austin published a book called *Climatology*. As in Rossby's case, the authors were not card-carrying climatologists. Haurwitz was a dynamic meteorologist and Austin a synoptic meteorologist. They felt it worthy of more than 40% of the space to discuss component weather phenomenology, such as air masses, fronts, cyclones and anticyclones, and also climatic mechanisms such as elements of the heat balance, the effects of land and sea (including the complexities of terrain and ocean currents), and even the wind and pressure distribution at higher altitudes.

Well, how will all this tie into the man we have come to honor? Landsberg's career spanned this very rapid transition in the field of climatology. In addition to his book *Physical Climatology* (L1958-c), there were other indications of his growing preoccupation with the structure and nature of climate and climate change, the reasons and the mechanisms, and man's possible influence. It was evident from some of the articles he began to write in the 1940s and extended right to the 1980s.

In a contribution to the *Handbook of Meteorology* (L1945-a), Landsberg wrote a remarkably well-rounded article, pointing out in his introduction that "the climatic conditions can be derived from a study of the general atmospheric circulation and its local modifications". He discussed the planetary radiation regime, the general circulation and continental, oceanic and topographic influences. He also touched on the more localized influences of lakes, of vegetation and human activity, and of regional dynamics. All of this was nicely summarized in a figure (Figure 1). It was very good for its day, and would even be useful today.

In contrast to this Handbook article, Landsberg's 1958 edition of *Physical Climatology* was less exciting, but he purposely confined its scope. However, in doing so, he made some interesting observations. He explained that "physical climatology" is a branch of the total field, complementary to descriptive or regional climatology. He noted, however, that:

> A further branch, dynamic climatology, closely allied to physical climatology, has emerged in recent years. It rightly recognizes that some of the basic facts of climate, as observed on the globe, stem from, and can be explained in terms of, the general atmospheric circulation. The features of this circulation, its repetitive tendencies, its flow patterns, eddy frequencies, its transport of heat and momentum are of fundamental importance for the theory of climate. Within our present framework only a few fleeting references can be made to this rapidly developing field."

This conclusion was qualitatively consistent with Landsberg's perceptions in the 1940s, as we have seen.

However, quantitatively, this allusion was a correct reflection of developments in the 1950s. And it was exemplified by two contemporary conferences, both organized by Bob White, who was then with the U. S. Air Force's Geophysics Research Directorate. One was a meeting on the Application of Numerical Integration Techniques to the Problem of the General Circulation held at the Institute for Advanced Study at Princeton in 1955. The other was a symposium on Long Range Forecasting held in August 1956 at the University of Wisconsin in Madison. I think that some of those who participated are here today. What was startling was that a new breed of young turks, never before connected with the field of climatology, were beginning to implement some of Rossby's strategy. What seems natural and plausible today, must have struck the climatologists then as audacious. But, evidently, not Helmut Landsberg. He might even have graced this new bunch by calling them "climatologists".

Much of the definitive and, in fact, decisive empirical work was to come as a result of the greatly enhanced war-time aerological network and the remarkable analyses by Victor Starr and Jakob Bjerknes and their associates in the 1950s and also in the 1960s. The theoretical and experimental capability would have its beginnings rooted in the methodology of numerical weather prediction, starting in the mid-1950s. Specifically, it began with the breakthrough in energetically self-sustaining long-term model simulations achieved by Norman Phillips. The term "general circulation model" was coined, as was "numerical experimentation". Rossby was fortunate to have witnessed this threshold in the realization of his vision.

What Rossby didn't live to see was the rapidity with which the fields of climate dynamics and long range forecasting would develop in the following three decades. Of course, we still have many unsolved problems, but the strides have been astonishing.

Mathematical modeling and simulation have completely transformed meteorology and climatology. They have enabled a precise and consistent accounting of the non-linear interactions among dynamics, thermodynamics, moist processes, radiative transfer and dissipative processes under the influence of general boundary conditions and from arbitrary initial conditions. The synergistic interplay between the components of an integrated geophysical system: the atmosphere, the oceans, the continents, the cryosphere, and, ultimately, the biosphere, can be studied. Predictability properties of time-dependent systems can be investigated and defined. The experimental approach, emulating controlled conditions in the laboratory, can be applied to study the interaction and relative roles of participating mechanisms and to systematically test hypotheses. It is Rossby's wish list and a lot more. But then again 45 years have now elapsed.

Moreover, the transformation has demanded an active embracing of many of the involved sister disciplines.

Oceanography became an integral part of the climate problem; some oceanographers might even claim the most important part. New demands were placed on our understanding of sea ice dynamics and thermodynamics.

Glaciology has become important not only as an indicator of past climatic variation, but as an independent means to determine the past composition of the atmosphere; for example, carbon dioxide and dust.

Atmospheric chemistry and its interaction with the chemistry of the other components of the climate system, namely the complex biogeochemical cycle, has become a major new discipline, bearing on inert pollution problems as well as highly interactive processes, such as ozone photo-chemistry and the greenhouse effect. The physics of the life cycle of a variety of aerosols and their optical properties constitutes another rapidly growing climate-related discipline.

The biological interactions with climate, as well as the biological reactions to climate, constitute a whole new interdisciplinary involvement. The full consequence of human activity has yet to be understood.

The field of geology also is growing rapidly in its direct relevance to building substantial scientific foundations in our understanding of climate variations. Paleoclimatic reconstructions are key to establishing the structure of past climatic extremes, and among other things, have been playing a unique and vital role in validating climate models.

Over the years, Landsberg left ample evidence of his keen appreciation for the growing and essential role of this much broader scientific context for the field of climatology.

Perhaps one of the most revealing recent perspectives was contained in an address he gave in May 1983 to the IX World Meteorological Congress in Geneva. Its title was "The Value and Challenge of Climatic Predictions." As usual, it was a consummately scholarly piece (L1985-e).

After reminding his audience of the socio-economic value of climatic information, he devoted most of his time assessing the current methodologies for prediction, both statistical and physical. It doesn't surprise me that Landsberg did not feel uneasy about merging the fuzzy transition between weather and climate.

In this talk, he reasserted his confidence in the physical approach. He judged "the fact that the atmosphere is a physical system, albeit a very complex one and subjected to multiple force influences, permits us to maintain a hopeful attitude toward predictability". He was willing to accept, what he referred to as, the "optimistic stance of the meteorological community" that climate was, in principle, predictable. However, Landsberg was not ready to accept the current status of our understanding as sufficiently useful. Let me quote a paragraph:

> If forewarned is forearmed, the meteorologist is not yet prepared to furnish the needed predictions. Yet planners, emergency services, and high-level government administrators are pressing for instantaneous, and simple, answers. They want to know where and when droughts will occur, they want projections of extreme heat or cold, they want pronouncements on occurrence of storm surges. Their responsibilities for safety and for the supplies of food, water, fuel, and power have made them very weather conscious. Thus they cajole and pressure the meteorologist and climatologist for projections. Often such insistence causes 'hasty' answers and the recipient of a long-range forecast or climatic estimate frequently forgets about the uncertainties and disregards the cautionary terms in which the answers are usually couched.

This problem, the pressures from the user community and temptations of easy virtue on the part of some meteorologists, is one that Landsberg discussed with me on several occasions. It was one of his deep concerns. To me, it was an eloquent measure of the man's integrity.

Landsberg felt, in the meantime, that "pedestrian statistical analog procedures will have to serve. Their usefulness should not be underrated and considerable improvements can confidently be expected."

He judged, however, that "much more difficult to achieve will be longer term climatic projections in quantitative terms. There is no complete physical theory of climate and the complexity of the feedback mechanisms will plague us for some time to come." He appreciated the potential of understanding and modeling El Niño and the Southern Oscillation and of anthropogenic effects. He understood the essential interactive role of the oceans. He worried about the limitations of the unpredictability of volcanic eruptions and of the limited predictability of solar events, which was also evident from some of his earlier writings.

Landsberg welcomed the rapid evolution (or perhaps even revolution), and understood and appreciated how it would expand our understanding of the physical basis of climate and enhance the ability of the science to serve mankind.

Landsberg's list of scientific and technical publications contains something over 300 entries. His fantastically diverse interests, as evidenced in this written legacy to us, almost always returned to his abiding central concern: the application and impact of scientific knowledge to man's well being on this planet. And, perhaps, more than any other way, that is the way I will remember Helmut Landsberg.

References

Haurwitz, B. and J. M. Austin, 1944: *Climatology*. McGraw-Hill, 410pp.

Kendrew, W. C., 1942: *The Climates of the Continents*, 3d Edition. Oxford Press, 473pp.

Rossby, C. G., 1941: The Scientific Basis of Modern Meteorology. In *Climate and Man*. Yearbook of Agriculture. U.S. Government Printing Office, 599-655.

Who's Who in the World, 6th Edition, 1982-1983. Marquis, 624.

Commemorative Ceremony Remarks, February 10, 1986

As in everything else, Helmut Landsberg was an activist when it came to his professional affiliations.

When the American Meteorological Society was barely 16 years old, Helmut Landsberg had already become a member. It was the beginning of a very active participation which was to continue until his death.

Over the years, Helmut served on many of the Society's specialized committees as well as on its committees of governance. He was involved in the editorial oversight of several of the Society's publications. He was elected to two terms as Councilor and in the early 60s was elected vice-president of the Society, receiving the largest number of votes ever recorded for that office or for that of the presidency itself.

As in everything else that Helmut Landsberg undertook in his long and fruitful career, his service to the Society was dedicated, sustained and distinguished.

His extraordinarily outstanding professional contributions have been acknowledged by the American Meteorological Society with honor on several occasions: for Outstanding Achievement in Bioclimatology in 1964; elected Fellow in 1968; the Charles Franklin Brooks Award for Outstanding Services to the Society in 1972; and the Cleveland Abbe Award for Distinguished Service to Atmospheric Sciences by an Individual in 1983.

Last summer, action was started within the Society to elect Helmut to Honorary membership, an esteem accorded only 12 times since 1960. Honorary membership is meant to delineate and to acknowledge an eminence comparable to or exceeding that demanded by the Society's highest awards. It should be for accumulated distinction over a number of years. The Council has agreed that "the addition of a new name to the roster of Honorary Members should be as much an honor for the Society as it is for the individual".

The timing of the presentation of his certificate of honorary membership had originally been intended for Helmut's birthday celebration. We are pleased and grateful that Frances Landsberg can honor us with its acceptance at this time.

The certificate text was published in *Bulletin American Meteorological Society*, **67**(12), 1498.

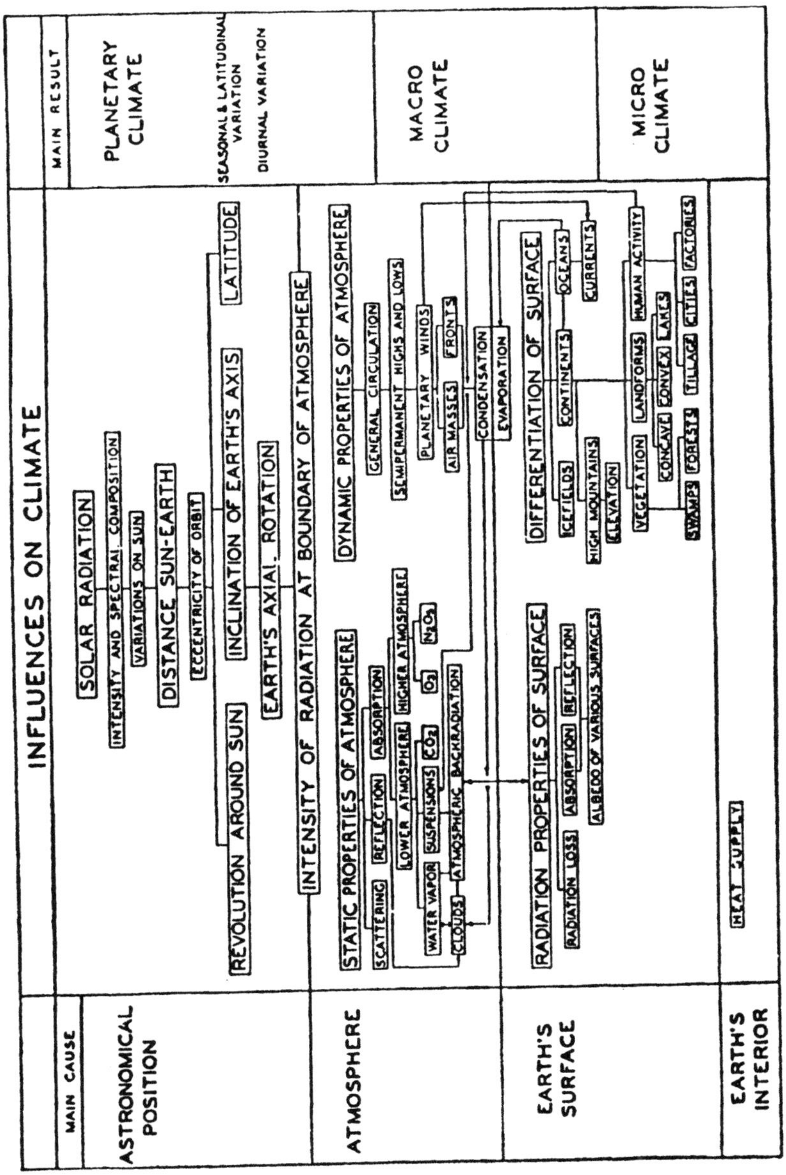

Figure 1. Scheme of influences on climate.

FIVE THEMES ON OUR CHANGING CLIMATE

J. Murray Mitchell

Introduction

A scholar of truly peripatetic interests, Helmut Landsberg had become attracted early in his career to the mystery surrounding climatic history and change. Toward the end of his life he devoted more attention to this topic than to any other.

"Fluctuations of climate ... are among the most fascinating problems in climatology," Landsberg wrote in 1958, in a perceptive and wide-ranging survey of the science he would do much in the course of his career to enrich and vitalize (L1958-e). Like a moth attracted to an irresistible flame, he returned again and again to the topic of climatic change and variability for scrutiny from a wide range of perspectives. His interests in.the topic were threefold: What are the facts of climate variability? How are the facts to be explained? How should society accommodate itself to climatic variability?

Whenever his many duties and responsibilities permitted, Landsberg absorbed himself in a variety of research thrusts into such questions. His broad education and disciplined mind, coupled with what seemed to his colleagues a boundless store of intellectual energy and curiosity, led to a steady flow of scholarly publications, essays, and commentary. A voracious multilingual reader of the literature from around the world, he was quick to recognize the significance of new ideas and new research tools. Some of these he enthusiastically adapted in new research thrusts of his own. Others, his intimate knowledge of the history of the science led him to unearth similar ideas already enunciated generations earlier, published in obscure and long-forgotten works he delighted in resurrecting.

So daunting would be the task of distilling all of Landsberg's many contributions to this field that I briefly thought of organizing this chapter along the lines of an encyclopedia, with some three to four dozen entries. I decided instead to limit myself here to identifying what I see as five main themes in Landsberg's prolific writings in this field. In each of these theme areas I shall focus on selected works of his that illustrate rather than delimit the range of his interests and scholarship.

Starting with the concept of "climate as risk," I call attention to Landsberg's keen interest in the application of climatic data in addressing a broad range of problems related to human comfort, health, recreation, and resource management. While director of the U. S. Weather Bureau's Office of Climatology, he promoted a major effort to exploit newly developed statistical approaches to the estimation of probabilities and extreme-value functions, as applied to the enormous store of climate statistics at his disposal. His objective was, first, to determine what kinds of "forecasts without date"—as he called them—would best meet the needs of planners and designers in such areas as agriculture, engineering, and water-system management; and, second, to provide such information in suitable data summaries and publications.

F. Baer et al. (eds.), Climate in Human Perpsective, 37–59.

Continuing with the theme of "climate rhythms," Landsberg had always been fascinated by historically rooted lore to the effect that climate varied in cycles and "rhythms." It was typical of him to approach such notions with respect and an open mind, but to involve himself and his colleagues in efforts to verify their credibility with modern data and statistical tools. Only when they passed muster in these respects was he willing to embrace them as real features of climate. This led him, at various times in his long career, to enquire into many alleged climatic rhythms, among them the quasi-biennial oscillation, various solar cycles, the so-called Brückner cycle, and the El Niño-Southern Oscillation phenomenon. While he came to accept some of these as probably real, by no means did he embrace all of them as bona fide rhythms (least of all, the Brückner cycle). He never pretended that such rhythms were the key to climate prediction, only a clue to the workings of the atmosphere. Indeed, he cast a disparaging eye on those who promoted the simplistic use of cycles, solar or otherwise, as a reliable forecasting tool.

It was on the theme of "climatic trends" that Landsberg addressed some of his earliest research in climatology. Already by 1951 he had examined the long temperature record at New Haven and a 60-year series of cloudiness observations at Washington D.C., and had found trends in both which he ascribed to a change of large-scale atmospheric circulation that had decreased the continentality of the climate in the northeastern U.S. as well as in Europe. In later years, however, by which time such trends had reversed direction at least once, Landsberg abandoned the term "trends" as a natural phenomenon, and spoke instead of "fluctuations." This was consistent with his view of climatic change as a phenomenon largely stochastic in character, and driven by a multiplicity of causal agents and climate-system feedbacks that he believed will require a very long road of future research to unravel. He was more pessimistic than many of his colleagues about the prospects for useful climate predictions to come out of climate-system models as "incomplete" in physical content as those now in use or on the horizon.

A theme that received a great deal of Landsberg's attention over the years was in the area of "human influences" on climate. Although he is likely to be remembered best for his classic studies of urban effects on local climate, Landsberg addressed a far wider range of human influences—both real and alleged—in many of his writings. His interest in this subject was stimulated by two circumstances. On the one hand, he was bothered by irresponsible press accounts—most of which appeared in the 1950s and 1960s—of how human activities of various kinds were said to be drastically changing the weather, and he felt an obligation to set the public record straight with reasoned analyses of his own. On the other hand, he was aware of claims of "one-way" global-scale climatic trends taking place, and attributed by some of his most respected colleagues to rising air pollution levels. He responded by gathering some of the longest weather records available around the world, and showing that they lacked what he saw as clear evidence of one-way trends. He maintained that this showed the climate system to be so robust that it was relatively immune to global-scale human influences, at least to the present time. Looking to the future, however, he often spoke of the CO_2 "greenhouse" issue as a major long-range concern that merited attention. But during the last years of his life he increasingly distanced himself from those of his peers who expressed confidence in the results of CO_2-related and other modeling studies as a reliable guide to future climate developments.

Finally, Landsberg had a special fascination for what I call the "solar factor." During his long career he devoted much time and energy to trying to sort out the natural physical origins of climatic variability and change. Early on, he recognized the potential but unproven roles of such environmental forcing as solar variability and volcanic effects. In his later years he often stressed the need also for a better understanding of the oceans and their role in producing longer-term stochastic fluctuations of climate. While he viewed climatic fluctuations, overall, as a collective response to a range of such forcing influences, he seemed to believe that solar forcing in particular was less understood than the others and deserved a more respectful look than most of his peers were inclined to give it. By that he meant the possibility of variations in the energy output of the sun, not only on the time scale of the 11-year sunspot cycle but over longer periods as well. Perhaps more than any other of Landsberg's eclectic interests in climatology, the "solar factor" absorbed him during the last years of his life.

Climate as Risk

FORECASTS 'WITHOUT DATE.'

Human society is ceaselessly buffeted by episodes of heat, cold, flood, drought, storm, and wind. The ideal response would be for us to apply our technology to control the elements. The next best response would be to predict the vagaries of the atmosphere in time to take protective action. As Landsberg was well aware, and often reminded us in his writings, often our ability to predict nature is almost as limited as our ability to control it.

Our best recourse in the circumstances, Landsberg maintained, is to quantify the range of atmospheric impacts on us in terms of risk. As he saw it, the assessment of risk can be regarded as a forecast "without date," with a value to society all too often overlooked. He viewed our long heritage of weather observations as a largely untapped goldmine of information available for assessing risk. In a nation like the United States, where most people are well acquainted with the notion of risk in areas of daily life such as investments, games, sports, and hazardous working conditions, he sometimes marveled aloud that so many of us seemed unprepared to look at our relationship to the vagaries of nature in the same way.

In the pages of Landsberg's earliest treatises in the area of climatology, dating back as far as the first edition of his classic textbook on physical climatology (L1941-f), one can find many passages in which he alluded to the value of looking at hazardous weather in terms of probabilities. Later on, as director of the Weather Bureau's Office of Climatology, he promoted the collaboration of specialists in extreme-value theory to develop climatologies of rare events—among them maps of the frequency of maximum winds, and of snow loading, needed in the design of buildings and other engineering structures.

Perhaps the most important of Landsberg's contributions in this field was a report on geophysical predictions, prepared under his leadership by a panel of the National Research Council's Geophysics Study Committee (L1978-b). This dealt with the entire range of geophysical disturbances of concern to society, from earthquakes, volcanoes, and landslides; to solar-related magnetic disturbances, weather and climate prediction, and severe storms; to tsunamis, tides, storm surges, stream flow, and flooding. The following passages are excerpted from Landsberg's masterful overview chapter in the 1978 report.

> Many volumes could be filled with descriptions of these scourges of mankind. Timely warnings (of them) will reduce loss of life, but they seldom avert the destruction of property. A good example in recent memory is the havoc wrought by hurricane Agnes and its accompanying flash floods. In four days in June 1972 it killed 118 people and caused $3.5 billion in property damages, a record for losses of private and public property and facilities.
>
> ... (P)rediction of future events is a major aim of all geophysical sciences. ...In no other field of science is prediction so intimately related to human safety and welfare, because geophysical events are among the greatest threats to people. The populations exposed to violent geophysical hazards are large.
>
> In many geophysical disciplines the curiosity of our ancestors about natural phenomena, the faithfulness of chroniclers, and devotion to duty of scientific observers are of unique value. The collection of further data and their safe storage in data banks are indispensable insurance for future improvements in geophysical predictions. For example, risk assessments of earthquakes and of volcanic eruptions, and decisions on land use, owe much to seismicity maps and (to) charts of past volcanic activity.
>
> All warnings of impending disasters ... require the closest cooperation of the geophysical (forecaster) with other agencies and with the people potentially affected. Here important tasks of communication and education have to be faced. Prediction alone is not complete protection for those in potential danger. Adequate and timely communication of such predictions is essential. These have to be followed by appropriate and timely action. ... Disaster education ... must include not only instruction about the potential danger of geophysical hazards but also guidance on what to do when such dangers seem to be imminent.
>
> ... (T)here is considerable value in predictions where (the time of an event) remains open. Such predictions without time or date are continually used in planning or design. The knowledge that in a particular area a geophysical hazard exists will dictate precautions to minimize potential loss of life and damage to property.
>
> For the design of such structures as dams and bridges, hydrological predictions of maximum floods for a given time interval, such as a 50-year estimated lifetime, are made. ... Thus technically one speaks of a 50-year, or even a 100-year, flood. This is often (mis)interpreted to mean that such flood-threatening storms occur at about 50- to 100-year intervals. But that is not the meaning of probabilistic statements. It is quite possible that storms expected to occur only once in 50 years might take place in successive years. The probability of such a happening may be small, but it is finite. For example, Kansas City, Missouri, recently experienced two separate and distinct rainfall events in 24 hours, each of which exceeded the 100-year rainfall frequency.
>
> ... (P)redictions without date, aside from their value for setting insurance rates, are of direct importance to planners, engineers, and architects. ... Where high winds are common, structural designs guard against (their damaging effects).
>
> Because of the rarity of such events, it is often difficult to resist the pressures for unsuitable development of land. There is no better example (of) this than the rapid and dangerous build(up) of developments along certain parts of the east coast of the United States where hurricanes are common or possible. In some areas they are sufficiently rare to lull the public into complacency. A typical example was the hurricane of 1938 along the New England coast. It had been over a hundred years since the last hurricane had struck

> there. In spite of warnings, people could not believe that such a storm was possible and did not evacuate. ... (Many of them) perished.
>
> In the approach to geophysical predictions ... there is a wide discrepancy in the understanding of causal relations and the actual capability to predict. ... There are reasons for this wide disparity. Public demand has played a major role (e.g., to the benefit of short-range weather forecasting and to the detriment of such obscure fields as solar variability). In (some) areas ... the basic observational tools and density of observations are quite inadequate to tackle more than the most general questions. Additionally, we cannot overlook the fashions prevailing in scientific research, which divert scientists into specific 'glamorous' areas. The total talent available to explore the (more) neglected subjects is about as scarce as the funding. ... Other areas that are in dire need of much closer examination deal with the social, economic, and legal consequences of predictions (and with the) proper communication of predictions to those potentially affected (L1978-b).

PREDICTION AS RISK

An aspect of risk not always sufficiently considered as such is concerned with the uncertainties inherent in weather and climate predictions. Landsberg addressed this subject on many occasions, usually in connection with long-range climate projections ventured from time to time by his peers. Here, I do not refer to his outspoken disdain for—and condemnation of—climate predictions based on naive physical reasoning, of the sensational kind that often made newspaper headlines and that he summarily rejected as irresponsible. Rather, I refer to his concern about long-range projections as issued by fully credentialed scientists, sometimes only in reluctant response to urgent public pressure for information, that he saw as vulnerable to public over-confidence in the projections. In his later years, he was inclined to criticize many public statements about the CO_2/climate issue in the light of such concerns.

In an invited scientific lecture for the Ninth World Meteorological Congress, in Geneva, Landsberg addressed what he called "The Value and Challenge of Climatic Predictions" (L1985-e). The following excerpts from that lecture illustrate the blend of optimism and pessimism with which he viewed various aspects of the climate prediction problem, in relation to the pressing public need for better climate forecasts.

> ... (T)he purpose of climatic information and predictions is to enable individuals and governments to cope with the vagaries (of climate). ... If climatic forecasts are to reduce the socio-economic effects of climatic variability, a number of ancillary steps are necessary: the weather-sensitive economic sectors must be identified, the flexibility of these sectors with respect to the forecast information must be documented, and then the relevant forecast capabilities have to be developed.
>
> The agricultural sector has long been identified ... as advantageously influenced by seasonal forecasts. The forecasts can be used in a decision analysis approach, and ... even relatively uncertain climatic information can be of some value (to agriculture) even though the skill levels of present seasonal forecasts are far from perfect.
>
> Another economic sector where even relatively short term projections can be highly useful is energy consumption. A case of recent climatic history is illustrative. The winter of 1976/77 was very cold in the midwestern U.S.A. The seasonal forecast did not include the most severely affected area in its below-average (temperature) prediction: it was labelled "indefinite." Natural gas and other fuel supplies were short in the most afflicted state, Ohio,

> and economic damages were estimated in the billions of dollars. A correct forecast, if heeded, could have placed fuel supplies in the right places at the right time. Aside from the fact that this exposed the shaky state of the (forecasting) art, it also illustrated the problem of prediction of extremes. ... All studies of climatic impacts reveal that (most major) disasters often deal with extreme weather situations. The flash floods, the severe local storms, the violent tropical storms are hard (enough) to predict on a day-to-day basis. How can one foretell that a season or a year will be afflicted by such menacing weather conditions?
>
> (T)he meteorologist is not yet prepared to furnish the needed (accuracy in) predictions. Yet planners, emergency services, high-level government administrators are pressing for immediate -- and simple -- answers. They want to know where and when droughts will occur, they want projections of extreme heat or cold, they want pronouncements on occurrence of storm surges. Their responsibilities for safety and for the supplies of food, water, fuel, and power have made them very weather conscious. Thus they cajole and pressure the meteorologist and climatologist for projections. Often such insistence (promotes) "hasty" answers and the recipient of a long-range (projection) frequently forgets about the uncertainties and disregards the cautionary terms in which (it) is usually couched. Quite often such (information is expressed as) an estimate of probability of occurrence, e.g., 'the chances are 3:1 that a particular weather condition will occur.' Unfortunately many decision makers and the public in general are not too certain about the proper interpretation of the language of probability.
>
> We must now ask, is the climatic future indeed predictable? The optimistic stance of the meteorological community is essentially affirmative. (L1985-e.)

At this point, Landsberg quoted from papers of the 1979 World Climate Conference maintaining that beyond "a week or so, there is a reasonable expectation that some features of the (climate) variability may be predictable for much longer, e.g., the tracks of cyclones, the incidence of blocking, some aspects of the monsoon, etc." (WMO, 1979)

Among non-meteorologists the opinion on predictability even of seasonal weather is less optimistic. ... But whether one is optimistic or pessimistic with regard to future developments of climate forecasts, it must be admitted that (such forecasts) are presently inadequate. (L1985-e.)

Citing Wiin-Nielsen (1976), who maintains that present-generation general circulation models show promise for allowing weather predictions out to two weeks, Landsberg quoted him further as a guide to "where we should go from here:" "The predictability problem on the climatic time scale requires a totally new approach. Direct 'brute force' integrations are likely to be replaced by a stochastic approach in which slow changes of climate are described as the response to continuous excitations by the shorter period disturbances and changes in the external forcing."

The fact that the atmosphere is a physical system," Landsberg continued, "albeit a very complex one and subject to multiple forcing influences, permits us to maintain a hopeful attitude toward predictability." In his inimitable style, he then devoted the body of his lecture to an enlightening and well-referenced review of the history of long-range forecasting and of the many concepts both physical and empirical that underlie present-day methodologies in the field. Closing his review by focussing on the role of the oceans in longer-range climate processes, Landsberg wrote:

> All of this points up the (need) to integrate oceanic conditions into all realistic schemes of long-range forecasting. ... Our present knowledge of the deeper ocean circulation and its influence on the upper layers is so inadequate that I question the validity of projections from current climate models for years and decades ahead. Here lies one of the major challenges for the future (L1985-e).

Climate Rhythms

WEATHER SPELLS

At one end of the scale of Landsberg's wide-ranging contributions to the study of regularities in weather and climate was a commentary he wrote on weather "spells" and "singularities" (L1960-f). The paper is notable for the physical as well as historical perspective it lends to the notion of weather "spells." The term *spells* refers to a tendency in some locations for the normal seasonal march of weather to evolve irregularly rather than smoothly from week to week, in such a way that similar weather anomalies are inclined to repeat near the same calendar dates in different years. *Singularities* is another term for the same phenomenon, but one usually defined in the more restrictive—and to Landsberg's mind, unrealistic—sense that the same weather anomalies are associated with the same calendar date each year.

The motivation for the paper was the publication the year before by three colleagues (Barger, Shaw, and Dale (1959)) who tabulated rainfall probabilities by weeks of the year at 125 stations in 12 north-central states in the U.S. "In glancing through these tables it was quite notable that certain weeks of the year were more prone to show precipitation than others." The tables specifically showed a preference for precipitation to occur in early April and early June, and a tendency for dry spells to be favored in weeks around late October, mid-November, mid-December, and the end of the year.

Landsberg traced the origin of the concept of "spells" to A. Buchan, who in a textbook published in 1871 noted "interruptions" of the regular annual course of temperature in Scotland. Schmauss (1928), he added, "revived the idea that specific weather events could be tied to individual calendar days," which led to extensive scientific debate at the time. As to the plausibility of "spells" as a real physical phenomenon, Landsberg wrote:

> The general circulation ... is likely to show recurrences of the same patterns at about the same time of the year. In each area these patterns have their own (topographically induced) peculiarities. Moreover, these tendencies may be overshadowed in abnormal years by other events and the spells may not show up at all. If we regard them in this light they become a much more acceptable concept to most meteorologists (L1960-f).

While the phenomenon of spells (singularities) is still sometimes used as a tool for long-range forecasting, especially in European countries bordering the Alps, it has many detractors as a valid basis for adding forecast skill. In connection with forecasting, Landsberg had this to say of the rainfall probability tables of Barger, Shaw, and Dale:

> (T)he tabulations ... should be of some interest to forecasters. ... (But) some forecasters feel that it is not 'cricket' to use climatological information. ... Of course, no one wants to advocate the use of climatological statistics indiscriminately. The skill enters when the forecaster decides when to follow the climatic odds and when not to, but the forecaster should by all means look at the odds (L1960-f).

THE "BIENNIAL PULSE"

In 1961, the discovery of the Quasi-Biennial Oscillation (QBO), consisting of a systematic reversal of wind direction from one year to the next in the tropical stratosphere, was made by Reed et al.(1961) and by Veryard and Ebdon (1961). Within a year of that discovery, Landsberg was to publish a remarkable survey paper on "biennial pulses in the atmosphere" (L1962-b), that traced the history of evidence of quasi-biennial waves in climatic data back into the 19th century.

In his introduction, Landsberg showed his flair for putting his topic into perspective. Noting the "will-o'-the-wisp" character of alleged regularities in atmospheric changes, he commented:

> A fundamental belief exists among meteorologists that not all is turbulent disorder with respect to time. This belief has lead to a continued search for some natural harmonies in the disorderly march of weather. Many of the great and near-great in the profession have participated in this search.

The issue, he observed, has been "beclouded" for two reasons: "the notion of preconceived causes", and "the zeal to consider use of cycles for forecasting. In the former category," he went on to say, "it has probably not been helpful that researchers have endeavored too assiduously to find solar influences in the data."

With specific regard to the QBO, Landsberg wrote:

> In scanning the literature on rhythms in the atmosphere one is soon struck by the fact that only a few of the alleged existing cycles appear in many of the series that have been analyzed. Most prominent among these, and evidently quite widespread and remarkably persistent, is a period of slightly over two years. As far as I can ascertain, this was first discovered by H. H. Clayton in 1884 on the basis of American temperature data, taken at widely spread localities, covering the period from 1839 to 1882. He (initially) placed the length at 25 to 25+ months (L1962-b).

Beginning with the reference to Clayton (1884), Landsberg listed 39 determinations of climatic rhythms in the range 2 to 2+ years, by some 20 authors who had analyzed long series of weather statistics from a wide range of world locations, culminating with the (then) latest upper-air tropical wind studies of Reed et al. and Veryard and Ebdon. Recognizing that the "biennial pulse" always seemed to be a weak—albeit a persistent—cyclic signal in surface climatological data, Landsberg added this perspective:

> The most important realization resulting from the analyses of (cycles) was -- or one ought to say should have been -- the inutility of cycles for forecasting weather. ... Even the two-year cycle, which is without doubt real, contributes only a few percent of the variance.

> This should not discourage us from seeking the cause of this and other significant, persistent rhythms because they may give us important insight into the atmosphere's mode of operation (L1962-b).

A year later, the theme of the "biennial pulse" was expanded in a study that Landsberg undertook with three of his Weather Bureau colleagues, and published in a collection of contributions in memory of Harry Wexler (L1963-b). This study began by recognizing the fact that the quasi-biennial oscillation so prominently reflected in the brief record of tropical stratospheric winds had also been detected in a wide range of surface meteorological series. The phenomenon accounted for a very small fraction of the total variance in the surface data, and the application of newly developed power spectrum techniques was necessary to isolate the QBO signal reliably. That signal showed above the spectral noise level so consistently that the opportunity seemed at hand to investigate the continuity of the QBO in space and time, to determine its global-scale geographical coherence and the consistency with which the QBO was evident in decades and centuries prior to the modern upper-air record.

The strategy of the study was to select long series of monthly mean temperatures and barometric pressure for locations stretching along each of two meridians as widely as possible from the arctic toward the antarctic, and to apply a digital band-pass filter designed to reveal the history of changes of amplitude and phase of the QBO signal in each of the series.

The results indicated that a rather abrupt reversal in phase of the QBO occurred along both meridians as one moved out of the tropical Hadley regime and into extratropical latitudes, both north and south. They also suggested that the tropical and extratropical phases of the oscillation had each been similar along the two meridians for most—but not all—years since the start year of the analysis (1908). Large amplitude variations were found at all locations but these were not systematic from one location to another. Finally, evidence of phase locking of the QBO with the seasons was found at the northern extratropical locations, such that the largest QBO amplitudes there coincided with temperature and pressure extremes reached in the winter months.

Among the further conclusions of this study was one that reflects Landsberg's great interest in historical precedents:

> ... (S)ome of the earliest known references to the biennial pulse in temperature, made around the turn of the century by Clayton, Woeikof, and Wallén, followed an epoch in which the pulse sometimes reached exceptionally large amplitudes. The pulse may therefore have seemed a rather natural subject for investigation at that time (L1963-b).

INVESTIGATIONS INTO OTHER CLIMATIC "RHYTHMS."

Throughout his early career Landsberg had been intrigued by recurrent claims in the meteorological literature of the existence of cycles in weather, most of which were alleged to be solar in origin. While his reaction to such claims was one of instinctive skepticism, it was typical of him to reserve judgment as to their validity until opportunities came along to check into them for himself.

In the late 1950s, Blackman and Tukey developed their theory of power spectrum analysis to enable a rational evaluation of the structure of periodic or quasi-periodic signals in time series. Landsberg immediately recognized the importance of this work as the key to investigating alleged cycles in weather and climate. By 1958, he had involved Harold Crutcher, a member of his staff at the National Weather Records Center, in Asheville, North Carolina, in programming the Blackman-Tukey procedure onto the (then) primitive computer system installed at the Center, and in exploring means to apply it to climatic series. This led, in 1959, to publication of a rigorous study (L1959-b) of the long and homogeneity-tested temperature record at Woodstock, Maryland, as the prototype of a series of investigations into the extent of cyclic behavior in climatic data.

The 1959 Woodstock study was a pioneering undertaking in many ways. Hans Panofsky and his students at Pennsylvania State University had been the first to apply the new power spectrum technique to the study of meteorological data, in connection with the study of the structure of atmospheric turbulence. But this was the first time the technique was to be used in the investigation of longer-term climate cycles, or what Landsberg preferred to describe as climate "rhythms." Its application was to pose a formidable logistic challenge to Crutcher because the calculations required to compute the spectrum (by way of the auto-covariance function) taxed the computational facilities at Asheville to its limits and required sequences of carefully planned memory dumps, at intervals of hours of computation, to arrive at the end result. A few seconds on a modern personal computer is all it takes nowadays, even without the benefit of modern FFT methods available to speed the computation!

The results of the Woodstock analysis were summarized in several areas, as follows: (1) Short-period variations of temperature were often—but not consistently—found in several ranges of periods, mostly around 3, 5-7, and 15-25 days related to intervals between airmass changes. (2) Long-period variations included a prominent one near 2 years, arguably one near the solar cycle of 11 years, and others beyond 50 years that could not be resolved in reliable detail. No evidence was found of a 22-year solar cycle, or of the roughly 35-year Brückner cycle alleged to exist in European data. (3) When the spectra were evaluated for individual months of individual years, each month was described as bearing a unique spectral "signature," evidently related to the prevailing Rossby wave number. (4) Spectra of temperature were less random than those of precipitation. (5) Higher harmonics of the 22-year solar cycle, claimed by C. G. Abbot as the basis of his controversial long-range weather forecasting system, were found to account for no more of the total variance in the Woodstock record than that expected by chance in a random spectrum.

In following years, Landsberg frequently relied on spectral analyses to explore "rhythmic" components of climate, among them the "biennial pulse" in the studies mentioned earlier. Meanwhile, various of his colleagues became involved in the development of "red noise" models to define the appropriate null continuum for significance testing of spectral signals in climatic series (e.g. Gilman et al. (1963)), and in the use of cross-spectrum techniques to assess the coherency of climatic variations with presumed environmental forcing factors (e.g. L1966-c).

Later, at the University of Maryland, Landsberg teamed with R. E. Kaylor and with some of his students to examine other long meteorological series by means of spectral methods, in one case including the maximum entropy method (L1977-f). With Kaylor, he authored a brief contribution on this subject for the *Journal of Interdisciplinary Cycle Research* (L1976-e), with the primary aim of acquainting a wider range of researchers into natural "cyclic" phenomena with the more objective assessment of cycles made possible by spectral analysis. There, he emphasized that while rhythms such as the biennial atmospheric pulse were shown to be evident in a wide variety of climatic series, they did not account for enough variance to be useful in forecasting yet they merited study as a clue to the behavior of global atmospheric dynamics.

Climatic Trends

Shortly after the end of World War II, a number of studies mostly authored in Europe cited evidence of a general warming of the climate in the northern hemisphere since the 19th century. Involved at the time in planning for the nation's programs in geophysics and geography, in President Truman's Joint Research and Development Board, Landsberg became attracted to the subject and carried out a study of his own (L1949-a). In that early study he focussed on the New Haven, Connecticut temperature record, the longest such record available in the United States which had started 168 years before, in 1780.

The New Haven study was remarkable for its statistical rigor in documenting the evidence of a long-term warming trend there, and for its illumination of possible physical factors responsible for the warming. Recognizing that the growth of the city may have contributed in part to the warming, Landsberg focussed on two seasonal indices that he reasoned would be relatively insensitive to urbanization effects but helpful in showing whether changes of atmospheric circulation might be involved. One index he chose for the purpose was the annual temperature range, or the difference between the monthly temperatures of the warmest and coldest months each year, that reflects the continentality of a climate. The other index was the Kerner isodromic index, involving the difference between the mean temperatures in April and October, that reflects the "oceanicity" of a climate.

Landsberg showed by means of plots of 30-year overlapping means of the two indices that the New Haven climate had been trending toward a generally less continental (and more oceanic) condition up to the end of the record in 1948. Pointing out that this result was in agreement with conclusions reached for several central European and other locations, Landsberg observed that ...

> ... (F)or a wide area in the middle and high latitudes of the circum-Atlantic space a trend toward a more oceanic climate has been in evidence for the past century. The immediate cause for this change must have been an increase in the intensity of the zonal circulation. ... This increase in circulation in recent periods has been proposed by (A.) Wagner and (Richard) Scherhag. Their findings are supported by the studies of (Thor) Hesselberg and (B.J.) Birkeland on pressure changes over the North Atlantic. ... Both (Franz Baur) and (H.C.) Willett have emphasized the role of the general circulation upon the secular variations of climate (L1949-a).

While these kinds of inferences may seem rather mundane now, they were something quite new and unconventional at the time Landsberg was writing. He then went on to show by means of various statistical tests, applied to comparisons between four non-overlapping subsets of data, that the changes in continentality (oceanicity) in the New Haven record were systematically reflected as parallel shifts in all parts of the frequency distribution of the data. He concluded, "So even in a statistical sense we are quite justified to speak about a climatic change at New Haven."

Two years later, on the eve of Landsberg's move from Washington to take charge of the Air Force's Geophysics Research Directorate, near Boston, he authored a second paper that lent a quite different perspective to the climatic change he had demonstrated to exist at New Haven. This time, he chose to look at the record of cloudiness at Washington D.C. (L1951-e), in which he found what he interpreted as further evidence of the trend toward a more maritime climate along the eastern seaboard of the U.S.

In choosing cloudiness for this study, Landsberg commented:

> Naturally, most of the past investigations (of climatic change) have centered around the records of the best observed meteorological elements: temperature, pressure, and precipitation. Observations of other elements have not been widely used ... The reasons are that they are often omitted from summary publications, such as World Weather Records, or that they are considered inhomogeneous or unreliable. There has been perhaps a somewhat unwarranted suspicion about these other data available from the records of long-established stations.
>
> When the great monographic compilation of climatic data from the capital of the United States became available recently (Zoch, 1949), it seemed intriguing to see if some of these neglected pieces of information might contribute to the analysis of recent climatic changes. ... We have used some of these data for the interval 1891 to 1950. This sixty year span comprises just two 30-year intervals, which are two basic time lengths considered as satisfactory in climatic practice for the establishment of a 'normal' (L1951-e).

In this study, Landsberg showed clear evidence of a systematic increase in daytime cloudiness at Washington in the 60-year period of record, manifest as both an increase in 'cloudy' days by about 25 percent and a decrease in 'clear' days by about 35 percent. As differences between the two 30-year periods, he showed that the cloudiness increase occurred, without exception, in all calendar months of the year. The amount of the change "was far too large to be attributable to errors or methodology of observation. They are exactly what one would expect had the climate become more maritime." He went on to examine humidity records for Washington, to check the possibility that the cloudiness changes were a local effect of urban growth, concluding that there had been a decrease of humidity over the 60-year period, too small and in the wrong direction to explain the cloudiness effect.

Landsberg was subsequently to comment in passing, from time to time, on the phenomenon of climatic change in various essays he would write on the state of the science of climatology. In the early 1960s, while director of the Weather Bureau's Office of Climatology, he authored two studies specifically devoted to the subject.

In one of those later studies (L1960-e) Landsberg mapped the geographical pattern of net temperature and precipitation changes in the United States between two 25-year periods of record, based on data from 48 non-urban locations, one in each of the 48 contiguous states, that had been judged suitably homogeneous for the purpose. These comparative data, for the periods 1906-1930 and 1931-1955, had been published by the Weather Bureau a year or so before in revised issues of *Climates of the States*. Landsberg had selected these non-urban data because he was aware of the problem of trend biases in data from first-order Weather Bureau stations, owing to local urban growth effects, and wanted to be sure of avoiding the influence of these in his analysis.

In tables and maps, Landsberg showed evidence of a net warming of climate in the U.S. since 1906, that was shared by most of the 48 stations in every season of the year and especially in winter and summer (as distinct from spring and fall). Using a t-test to gauge statistical significance of his results he found the annual average warming to be significant at 33 of the 48 stations, at the 95-percent confidence level. Stratifying the data into three zones of latitude, he also showed that the net warming between the two 25-year periods increased with increasing latitude: "This is in line with the findings, in other regions of the northern hemisphere, that the warming has been most pronounced in the higher latitudes."

Landsberg also looked at the precipitation changes at the 48 rural stations. Noting these to have remained generally within 10 percent, and lacking statistical significance in all but a handful of cases, he concluded:

> ... (W)e see that there are two areas in the United States where a reduction in rainfall was prevalent during the period. The Great Plains, Rocky Mountains, and Southwest show this tendency. With two major droughts in the second interval, this is to be expected. The droughts of the mid-thirties and early fifties are well remembered. Actually it is more surprising that, on the whole, rainfall held up rather well in the second interval. ... It would seem quite reasonable to state that no major trend in the rainfall conditions of the contiguous States of the Union is noticeable (L1960-e).

In 1961, studies of global-scale trends of temperature were published almost simultaneously by Callendar (1961) of Great Britain, and by Mitchell (1961) who was a member of Landsberg's own staff at the time. These studies each corroborated the global scale of the warming of earlier decades of the 20th century but used quite different selections of data and analysis procedures in reaching their conclusions. Landsberg immediately took a strong personal interest in these studies, and teamed with Mitchell (L1961-d) to undertake a comparison of the magnitudes of global-scale warming as estimated by the two authors for different ranges of latitude. He also sought to present reasoning alternative to Callendar's with regard to the possible origin of the warming which, in Landsberg's view, was less certain to be related to the greenhouse effect of carbon dioxide than Callendar proposed.

Among Landsberg and Mitchell's conclusions were: (1) when the data of Callendar and of Mitchell were pooled for each of various latitude bands from the Arctic to the Antarctic, good agreement as to the magnitude of the global-scale warming was found, and (2) this agreement tended to support the view of Mitchell (and of Willett before him) that a relatively small number of stations (between 120 and 180) sufficed to establish the global-scale changes as reliably as the much larger number of stations used by Callendar. Also, (3) Callendar was correct in concluding that the presence of local urban effects in some of the data averaged together to determine the global-scale temperature changes was not a serious source of error in either analysis. Landsberg and Mitchell, however,

(4) took issue with Callendar's view that the relatively small net warming found in the southern hemisphere could be explained on the basis that most of the CO_2 inputs to the atmosphere had originated in northern latitudes and did not yet have time to mix into the southern hemisphere, to impose an equivalent "greenhouse" effect there. Other physical factors, they reasoned, could have accounted for the smaller warming evident in the southern latitudes, perhaps including those related to "solar and dust" effects rejected by Callendar. In his reply to Landsberg and Mitchell, Callendar (see L1961-d) acquiesced in most of these points but went on to propose that the lesser warming in the southern hemisphere might be related to "the thermal inertia of the southern oceans and ice," an idea that would gain many adherents in years to follow.

After leaving the Weather Bureau and while at the University of Maryland, Landsberg maintained an interest in the subject of global climatic change. In 1975, for example, he teamed with J. M. Albert (L1975-m) to undertake an interesting study, based on a comparison of J. Hann's world temperature chart using data prior to 1885 with modern WMO temperature "normals" for 1931-1960. In that study, Landsberg noted the care with which Hann had adjusted his early data, as by reducing temperatures to sealevel and by avoiding urban stations, and applied similar rules in his treatment of the modern "normals." Among the results of this comparison that Landsberg emphasized were (1) evidence of a "ring" of temperature rises of some 2°C north of about Latitude 50°N, together with a cooling by a similar amount in lower-latitude continental areas; and (2) evidence of a net warming since the 19th century in the United States with the sole exception of certain southeastern states.

Human Influences

In the 1950s, concerns were expressed by some climatologists that human activities may be in the process of disturbing the natural balance of climate on a global scale. These arose in part from indications of a global-scale climatic warming in progress since the 19th century, coupled with early evidence of a long-term buildup of carbon dioxide and other pollutants in the global atmosphere.

Such concerns arose from another quarter as well: a new public awareness of the awesome power of the atom, coupled with the fact of massive radioactivity releases into the atmosphere, occasioned by a series of dramatic nuclear weapons tests conducted in the atmosphere in the years following World War II. On the one hand, the atmospheric tests fuelled fears that the spread of radioactivity might somehow be responsible for various aberrations of global climate occurring at the time. Those fears, however, mercifully faded away with the radioactivity, as the climate continued thereafter to behave in the manner that a closer look at weather records showed had always been its custom -- which is to say, richly punctuated by 'aberrations' of one kind after another.

In the minds of other observers, however, the atmospheric tests kindled ideas of a very different sort. Among these were grandiose notions of our being able to harness nuclear energy to modify global-scale climate at will. Newspapers began to carry boldly worded claims of the dawn of a new era in environmental control, in which the atom would allow us to tame the powers of nature and to 'manage' our global climate by building around ourselves a kind of protective shell against the elements. Technology --

or so the argument went at the time -- could triumph over nature and do so at a cost well within our national or international means to pay.

Landsberg was to become an early skeptic of such fanciful dreaming. Armed with a knowledgeable appreciation of the enormous energy that drives the natural climate system, he spoke on many occasions of the inherent stability of global climate that he believed would be difficult for man's technology to upset either deliberately or inadvertently.

In an effort to maintain a much-needed voice of reason on the matter, Landsberg turned his considerable personal resources and energies to gaining a longer historical perspective on the stability of climate. He considered that global climate was constantly subject to the disturbing influences of a range of natural environmental forces that, in his view, had much greater opportunities than did human activities to alter climate but seemed limited in their ability to do so.

His approach was to let the record of past climates speak for itself in terms of what he saw as ample evidence of recurring fluctuations in climate but a lack of 'one-sided trends' that one would expect to find if human influences were a significant source of change.

FOCUS ON LONGEST RECORDS

Many times over the years Landsberg was to collect and examine early weather chronicles, from various quarters of the globe, in efforts to extend the modern record of climate into earlier centuries for a longer view of the character of climatic fluctuations. He drew attention to long records already published, including the series of temperatures extending back to the late 17th century for central England (Manley 1953) and for central Europe (Baur 1962). He added his own reconstruction of the temperature record from 1780 at New Haven (L1949-a) and that of the precipitation record from 1751 for the vicinity of Boston (L1967-d). These were not only the longest such records available in the United States but, as Landsberg believed, they were largely free of observational inhomogeneities that afflicted most records of the kind and made them unsuited for the purpose.

In a comparative analysis of such long records (L1967-d) Landsberg focussed on the lack of 'one-sided' trends, other than one evident in winter temperature at New Haven that he attributed in part to local urban growth and in part to decreasing continentality of the climate in the present century (but with evidence of a reversal since the 1940s). He drew attention to anomalous individual years and decades, noting changes in the relative frequency of cold years, and that of unusually wet or dry decades, from one century to another. He noted that, at New Haven, the coldest summer occurred in 1816 -- the famous 'year without a summer' widely attributed to the eruption of Tambora --but that in other seasons other years were the coldest. He also noted a tendency for summer temperatures to vary slightly with the phase of the 11-year sunspot cycle, akin to a similar tendency that he and Mitchell had recently reported to exist in the record for Central Park, New York City (L1966-c). In his concluding paragraph, Landsberg wrote:

> In individual years peculiar conditions can, of course, be conditioned by terrestrial effects, such as volcanic eruptions. ... The climate still fluctuates under the influence of natural

forces. The old New England records reflect no strong one-sided trend that could be attributed to man's influence. So far the atmosphere and its driving forces are too overwhelming for the inadvertent interference of human activity, but the situation will require continuous careful scrutiny. (L1967-d)

POTENTIAL FOR HUMAN ALTERATION OF CLIMATE

Landsberg rejected claims of human intervention in the natural climate system as the explanation of global-scale climatic trends in evidence to date. On the other hand, he wrote often of the influence of human activities on local and regional climate, and of the growing potential for human alteration of future climate including the global-scale.

Of special interest to Landsberg was the influence of towns on local climate and the role of urban development in giving rise to local-scale trends, together with their many implications for the quality of life of urban residents. His contributions in this area of climatology were so substantial and far-reaching that they are the subject of a separate chapter in this volume.

Over the years, Landsberg also authored many remarkably insightful reviews and essays in which he addressed a far wider spectrum of issues related to human influences on climate (L1970-c; L1972-c; L1973-b; L1974-a; L1974-e; L1975-a; L1975-i; L1976-d; L1976-h; L1978-f; L1979-c; L-1980-c; L1982-d; L1983-f). The earliest of these was a lead article in Science (L1970-c), in which he introduced the subject with the following comments:

> ...(T)he question "Has man inadvertently changed the global climate, or is he about to do so?" is quite legitimate. It has been widely discussed publicly—unfortunately with more zeal than insight. ... There have been dire predictions of imminent catastrophe by heat death, by another ice age, or by acute oxygen deprivation. The events foreseen in these contradictory prophesies will obviously not all come to pass at the same time, if they come to pass at all. It seems desirable to make an attempt to sort fact from fiction and separate substantive knowledge from speculation. (L1970-c.)

Appropriately he went on to say, "In order to assess man's influence, we must first take a look at nature's processes." These he succinctly summarized, concluding with this:

> The fact that we have not yet succeeded in disentangling all the cause-and-effect relations of natural climatic changes considerably complicates the analysis of possible man-made changes.

Following with a rigorously documented review of what was known at the time about various categories of influence on climate, ranging from ocean interactions, carbon dioxide, atmospheric dust particles, and local effects both urban and 'extraurban,' he touched substantively on many controversial topics of the time. Among these were cloud seeding, hail suppression, experiments to modify hurricanes, proposals to alter climate by spreading dust on arctic sea ice, and possible stratospheric effects of large fleets of supersonic airliners. Landsberg concluded with a range of observations and forward-looking proposals that seem no less appropriate to be stressed today—nearly twenty years later. Among them:

...(I)t appears that on the local scale man-made influences on climate are substantial but that on the global scale natural forces still prevail. Obviously this should not lead to complacency. The potential for anthropogenic changes of climate on a larger and even a global scale is real. At this stage activation of an adequate worldwide monitoring system to permit early assessment of these changes is urgent. This statement applies particularly to the surveillance of atmospheric composition and radiation balance at sites remote from concentrations of population.

...In my opinion, man-made aerosols, because of their optical properties and possible influences on cloud and precipitation processes, constitute a more acute problem than CO_2. Many of their effects are promptly reversible; hence, one should strive for elimination at the source. Over longer intervals, energy added to the atmosphere by (local) heat rejection and (by) CO_2 absorption remain a matter of concern. (L1970-c.)

The scope of Landsberg's vision with regard to the potential consequences of atmospheric pollution was unusual for his time. In testimony he gave to a committee of the U.S. House of Representatives, as early as 1975, he addressed many such consequences including what was then the publicly unknown issue of 'acid rain':

(A) not inconsiderable by-product of air pollution should not be overlooked. Although washout of pollutants by rain is a welcome self-cleaning process of the atmosphere, some of the substances thus removed acidify the rain. This, in turn, will contribute to soil acidity (which) can cause agricultural problems. The acids washed out by the rain may have originated hundreds of miles away as effluents of sulfur dioxide or oxides of nitrogen. (L1976-d.)

The Solar Factor

SUN, WEATHER, AND CLIMATE

A category of potential forcing of the climate system by natural mechanisms that always appealed to Landsberg as a legitime target of scientific inquiry is that of possible solar influences. As early as 1949 he had discussed this in connection with the problem of explaining the ice ages (L1949-b).

While solar variability was a theme to which he returned numerous times during his career, it would be a mistake to characterize his interest in it as a personal conviction that it must play an important role in causing climatic fluctuations. More accurately to say, he was fascinated by the arguments and counter-arguments offered by his peers as to the importance (or unimportance) of solar influences on weather and climate, some of which had been vigorously debated over the years by the greats and near-greats of the profession. His own view was that of an open-minded observer who craved a better understanding of the sun, in particular of the capacity of the 'solar constant' to vary, and who was willing to reserve judgment as to its role in climatic change until more facts in these regards were to become known.

The cautious approach to the subject that would be reflected throughout his later writings was evident already in 1949, when Landsberg alluded to the well-known Smithsonian 'solar constant' determinations of C. G. Abbot:

> Present-day evidence suggests that there are at least short-periodic changes in the extra-terrestrial intensity of solar radiation. This evidence, in part, is based on calculations which, in detail, are open to criticism. This applies especially to the determinations of the so-called 'solar constant' (Abbot, 1935). ...(S)ome of these might soon be checked by more direct measurements taken at great altitudes from rockets. The observations of the past years indicate, however, that solar radiation actually fluctuates. It is less clear whether such fluctuations are anything but short-periodic, and it will take many years of observation before any long-term trends can be discovered. (L1949-b).

It was not until the 1980s, of course, that the reality of day-to-day 'solar constant' micro-fluctuations, of order 0.1%, was ultimately confirmed beyond all doubt by satellite measurements—albeit smaller in magnitude than those deduced by Abbot. At this writing, the existence also of a systematic longer-period fluctuation, again of order 0.1% and apparently in phase with the 11-year sunspot cycle, is strongly suggested by the satellite data (e.g., Willson and Hudson 1988; Livingston et al. 1988). Had he lived to read of this new information, Landsberg surely would have viewed it as a vindication of his career-long interest in sun/climate relationships, but no doubt he would have pointed to the small magnitude of the irradiance fluctuations as a problem in justifying claims of significant atmospheric responses to solar variability.

Perhaps the most comprehensive of Landsberg's assessments of the field of sun/weather relationships is one he wrote in 1983, in the pages of the *National Weather Digest* (L1983-g). He began it with the following remarks:

> Probably no subfield of meteorology has had as much effort devoted to it as the effects of solar variability on weather and climate. And none has had as little to show for the research labor. ... About a quarter century ago, *Meteorological Abstracts and Bibliography* compiled a literature survey which up to that time listed 1278 titles (Nupen and Kageorge 1958). ... A review paper by Pittock (1978) added about 150 more references written in the 1960s and 1970s.
>
> How did it all start? Although the existence of sunspots was known in classical antiquity, scientific observations of (them) began with the invention of the telescope by Galileo. In 1611 four early astronomers, including Galileo, announced their discovery almost simultaneously. ... The approximate (11-year) periodicity of solar activity was not published until 1844, and the first relation to meteorology was claimed by Koeppen in 1873 who cautiously suggested a negative correlation between sunspot number and temperature. That set the ball rolling and it hasn't stopped since (L1983-g).

Following an acknowledgment of the 'clinching' evidence from deep-sea cores of the influence of the Milankovitch orbital variations on climate, Landsberg went on to discuss the rocket and satellite determinations of the 'solar constant,' which he described as 'an undesirable term because it isn't constant.' He then summarized the situation, as he saw it, with regard to the purported evidence of effects of solar variations on climate and weather, respectively.

Calling attention to the irregular nature of the 11-year cycle, to the 22-year magnetic 'Hale' cycle, and to the 'Gleissberg' cycle (of order 90 to 100 years), he went on to say:

> The search for these solar periodicities in time series of meteorological elements, mainly temperature and (rainfall), has been the target for scores of studies. Most of them have been disappointing. ... (T)heir contribution to the total variance is usually less than 10 percent. (In) the longest temperature series available ... that of Manley for central England ... only about 6 percent of its variance could be explained by (the 11-year cycle). The power spectrum of a reconstruction of northern hemispheric temperatures since 1579 (L1979-f), about 400 years, fails to show any significant power for the shorter sunspot rhythms. However, a significant peak shows up for a 99-year cycle. This may be the Gleissberg cycle but with a series only four times as long, it would be bold to make any claims for its reality.
>
> Hoyt (1979) has suggested that the ratio of umbral to penumbral areas of sunspot(s), an indication of solar convection, would be a better index (than sunspot numbers) to indicate solar energy output. ... (He found that in the period between 1874 and 1970 for which he calculated such ratios) the correlation with northern hemispheric temperature values by Borsenkova et al. (1976) ... is 0.57.
>
> The method of Hoyt suggests that an active sun has lower energy emissions than when it is inactive. ... Papers by Eddy (1976, 1977) had postulated the opposite, namely that an active sun radiated more energy than a quiet sun. He based this in part on his impressions of effects of the so-called 'Maunder Minimum' of sunspots and its alleged cooling effects on the earth. This minimum of sunspots, placed in the interval of 1645-1715, while apparently low in solar activity, was not as minimal as represented. ... Also, the contention that the Maunder interval was particularly cold cannot be maintained. Actually (based on L1979-f, the) decades just before that period and toward the end of the 18th century and in the beginning of the 19th century were definitely colder in the northern hemisphere (L1983-g).

It should be interjected here that recent satellite information with regard to the 'solar constant,' referred to above, seems on its face more consistent with the stated view of Eddy than with that of Hoyt. Moreover, the reconstruction of northern hemispheric temperatures back to AD 1579 by Groveman and Landsberg (L1979-f), while a courageous undertaking, needs to be recognized as a highly tentative measure of the true history of hemispheric temperature, based as it was on geographically very sparse information prior to the mid 19th century. For that reason, one may be justified in viewing the Groveman-Landsberg reconstruction as an inadequate basis for drawing 'definite' conclusions about the relative coldness of the hemisphere in the Maunder Minimum period. To continue Landsberg's commentary:

> A better case can be made for (a climatic influence of solar activity rhythms on) precipitation. The power spectra for long precipitation series tend to show a notable coherence with sunspot numbers, especially in tropical and monsoonal areas. But even in this case the total ... variance explained remains small, usually below 10 percent. Droughts in the U.S. west of the Mississippi River have also been linked to the (22-year) solar rhythm. ... One is left with a rather general conclusion that droughts in the West will probably recur at about 20 year intervals with (an) uncertainty of 2 or 3 years in the timing (L1983-g).

He concluded with a wide-ranging survey of recent studies in which solar activity of various kinds had been implicated as a factor in short-period weather changes. These ranged from evidence of a connection of the passage by earth of solar magnetic sector boundaries to upper-tropospheric vorticity patterns, to evidence of a connection of solar flare-related modulations of cosmic radiation to invasions of stratospheric air into the troposphere, and to evidence of correlations between solar activity and electrical activity in thunderstorms. In each case he properly emphasized the tentative character of the evidence, the lack of adequate physical understanding of the alleged relationships, and the relatively weak character of the relationships that—even if verified as real—would be of little benefit to operational weather forecasting.

THE 'MAUNDER MINIMUM'

I return to the subject of the Maunder sunspot minimum, that occurred around the end of the 17th century, as a matter to which Landsberg devoted considerable attention in his last years. I do so partly to illustrate two of Landsberg's traits as a scientist that were highly valued by his peers. One of these traits was the great personal energy he devoted to tracking down and evaluating obscure historical documentation of natural phenomena that he considered relevant to current research investigations. The other was his abiding sense of fairness and intellectual honesty.

In 1980, Landsberg published a rejoinder to the claims of Eddy, to the effect that the Maunder Minimum was not as deficient in sunspots as Eddy maintained, particularly in the years after 1700 (L1980-f). Landsberg had laboriously waded through obscure manuscripts of sunspot observations from the 17th and 18th centuries to arrive at his position on the issue, including the journals of "the astronomer family Kirch, now scattered in various archives" where Landsberg had managed to unearth them. In the belief that the documents relied upon by Eddy had not taken the Kirch observations into account, Landsberg cited them in his 1980 paper, which he submitted as an invited contribution to an issue of the *Archiv für Meteorologie Geophysik und Bioklimatologie* dedicated to F. Steinhauser on the occasion of his 75th birthday.

Some time after his 1980 paper was published, Landsberg got wind of the possibility that Eddy's source documents, in particular the lists of sunspot observations between 1616 and 1715 assembled by Rudolf Wolf (1816-1893) of the Zurich astronomical observatory, may already have included the family Kirch observations. But no one known to Landsberg was able to confirm this. It was characteristic of Landsberg that he immediately began a hunt for the Wolf material. He located a typescript version of it, in the library of the Mount Wilson Observatory in California, only to discover that it lacked most of Wolf's commentary as to sources. So he undertook to track down a copy of the original handwritten version with the help of some of his many contacts in Europe.

"The German is in script but all of it is hard to read because of Wolf's abominable handwriting," Landsberg was to write when the manuscript eventually reached him. He, nevertheless, struggled with it long enough to manage to translate all but a handful of illegible words of it that pertained to the period 1645 to 1715—the inclusive dates of the Maunder Minimum as defined by Eddy. Here was a list of every individual sunspot along with a notation to indicate the source of Wolf's information. Sure enough, the

Kirch family was among a total of 20 astronomers cited as sources. Every spot that Landsberg knew to be recorded by that family had been included in the Wolf account used by Eddy.

By March of 1984, Landsberg had assembled a report detailing his findings and to which he appended his translation of the Wolf manuscript (L1984-c). In this report he wrote:

> The record does indeed show that the observations after 1700 indicate the solar observations of the Kirch couple in Berlin, although there are some minor irrelevant discrepancies in dates with the diaries. Hence, I withdraw all claims to have added to the data store in my 1980 paper.

He went on to supply a valuable graphic showing the yearly number of polar aurora sightings in the interval of the Maunder Minimum, based on about 100 auroral events in addition to those listed in the classic 1873 catalogue of H. Fritz, that Landsberg had culled from various sources. He did so in recognition of the fact that auroral frequencies could be taken as an independent measure of solar activity levels at the time, albeit not an infallible one. He regarded the auroral data he assembled as an indication that the sun was not as entirely inactive as the dearth of observed sunspots at the time might imply.

Other comments he made in the 1984 report reveal something of Landsberg's larger objective in pursuing studies of the kind. They also reveal something of the valuable perspective he might have contributed to the subject, had he lived to pursue his interests in it further. It seems fitting to close with a selection of those comments:

> There are now indications that (solar irradiance) is related to solar activity. This raises the hope that (information about this can be gained by means of statistical transfer function analysis) applied to the sunspot numbers. If this were successful, inferences of solar energy flux for over three and one half centuries might be possible. This would be of great importance to climatology. Hence the establishment of good series of sunspot observations and perhaps other indices of past solar activity is an important task.
>
> Much more evidence needs to be accumulated to 'calibrate' old sunspot observations. (Eddy and I) are in agreement that further search for information on 17th century observations is needed. There are still unexploited diary sources but practically all of them are from Europe which in the north and west is plagued by cloudiness. In southern Europe there may have been reluctance to note sunspot events after the ecclesiastical censure of Galileo in 1632 and (of) Scheiner. The notable exodus of Italian astronomers to other countries (e.g., France) makes further search not promising. There is more hope that further evidence of solar activity can be found in observations of aurorae. ... Practically every weather diary has reference to aurorae, even though in the 17th century one has to exercise great care in interpretation of Latin as well as vernacular terms. The label 'northern lights' did not come into common use until the 18th century. But still there are some unambiguous references (to aurorae).
>
> In reading Wolf's notes there are a number of contradictory statements. Some of the quotations refer to the dearth of (sun)spots in certain intervals; others indicate numbers as high as 50 in the Maunder interval. It is hard to decide if the original observers referred to the then still unknown and irregular sunspot rhythm or to absence of spots even when maxima should be observed.

Without question the Wolf notes are incomplete. One of the most curious omissions is the set of observations (by Bion) of October 1672 in Paris. If published sources (like Bion's) were not included, one may well conclude that other observational material in an age of poor communication may still exist. The tantalizing prospect of nearly 400 years of solar flux values for correlation with climatic data would make further search for observations very much worthwhile (L1984-c).

References

Abbot, C. G., 1935: Solar Radiation and Weather Studies. *Smithson. Misc. Coll.*, **94**, 135 pp.

Barger, G. L., R. H. Shaw, and R. F. Dale, 1959: *Chances of receiving selected amounts of precipitation in the north central region of the United States*. Iowa State Univ., Ames, 277 pp.

Baur, F., 1962: Langjährige Beobachtungsreihen. In: Linke (ed.), *Meteorol. Taschenbuch, neue aäsgabe*, 1, 710-803.

Bion, N., 1699: Les observations des taches du soleil. In: *L'Usage des Globes Céleste et Terrestre et des Spheres Suivant les Différens Systêmes du Monde*. Jacques Guerin(Paris), 345-348.

Borsenkova, I. I., K. V. Vinnikov, L. P. Spirina, and P. I. Stechnovskii, 1976: Variations in northern hemisphere air temperature in the period 1881-1975. (In Russian.) *Meteorologiya i Gidrologiya*, **7**, 27-35

Buchan, A., 1871: *Introductory Text-Book of Meteorology*. Edinburgh and London, 218 pp.

Callendar, G. S., 1961: Temperature fluctuations and trends over the earth. *Quart. J. Roy. Meteorol. Soc.*, **87**, 1-12.

Clayton, H. H., 1884: A lately discovered meteorological cycle. *Amer. Meteorol. J.*, **16**, 140, 183.

Eddy, J. A., 1976: The Maunder Minimum. *Science*, **192**, 1189-1202.

Eddy, J. A., 1977: Climate and the changing sun. *Climatic Change*, **1**, 173-190.

Fritz, H., 1873: *Verzeichnis beobachter Polarlichter*. Gerold's Sohn (Vienna), 255 pp.

Gilman, D. L., F. J. Fuglister, and J. M. Mitchell, 1963: On the power spectrum of 'red noise.' *J. Atmos. Sci.*, **20**, 182-184.

Hann, J., 1887: *Atlas der Meteorologie*. Berghaus' Physikalischer Atlas, Ab. III, Gotha, Justu Perthes, 12 pl., 61 figs.

Hoyt, D. V., 1979: Variations in sunspot structure and climate. *Climatic Change*, **2**, 79-92.

Koeppen, W., 1873: Über mehrjährige Perioden der Witterung. *Zeit. österr. Gestellshaft* f. Meteorol., 8, 257-272.

Livingston, W.C., L. Wallace, and O. R. White, 1988: Spectrum line intensity as a surrogate for solar irradiance variations. *Science*, **240**, 1765-1767.

Manley, G., 1953: The mean temperature of central England, 1698-1952. *Quart. J. Roy. Meteorol. Soc.*, **79**, 242-261.

Mitchell, J. M., 1961: Recent secular changes of global temperature. *Trans. N. Y. Acad. Sci.*, **95**, 235-250.

Nupen, W. and M. Kageorge, 1958: Bibliography on Solar-Weather Relationships. *Meteor. Abstracts and Bibliog.* Amer. Meteor. Soc. (Boston), 248 pp.

Pittock, A. B., 1978: A critical look at long-term sun-weather relationships. *Rev. Geophys. Space Phys.*, **16**, 400-420.

Reed, R. J., W. J. Campbell, L. A. Rasmussen, and D. G. Rogers, 1961: Evidence of a downward-propagating annual wind reversal in the equatorial stratosphere. *J. Geophys. Res.*, **66**, 813.

Schmauss, A., 1928: Regularitäten im järlichen Witterungsverlauf von München. *Dtsch. meteor. Jb. Bayern*, **50**, B 1-22.

Veryard, R. G. and R. A. Ebdon, 1961: Fluctuations in equatorial stratospheric winds.*Nature*, **187**, 791.

Wiin-Nielsen, A., 1976: Atmospheric predictability on various scales. *Die Naturwissenschaften*, **3**, 506-512.

Willson, R. C. and H. S. Hudson, 1988: Solar luminosity variations in solar cycle 21. *Nature*, **332**, 810-812.

World Meteorological Organization, 1979: *Proceedings of the World Climate Conference, Geneva, February 12-23, 1979.* WMO No. 537 (Geneva), p. 756.

Zoch, R. T., 1949: *The Climatic Handbook for Washington, D.C.* Tech. Paper No. 8, U. S. Weather Bureau (Washington, D.C.), 235 pp

CLIMATE OF CITIES

Timothy R. Oke

In a review of Helmut Landsberg's book *The Urban Climate* (L1981-d), I referred to him as "urban climatology's most respected doyen". The subject was enormously enhanced by him having lent his name and interest. During the thirty years of his most direct influence, he helped bring the field from a largely descriptive status to one with a much fuller physical understanding. But more than that he kept a much needed balance between research ideals and practical relevance through his gentle urgings and reminders.

In the first part of this paper, I will provide an historical overview of developments and personalities in urban climatology and Landsberg's relation to them. In the second part, I will illustrate his contributions by liberal reference to his own writings on the subject. Finally, we will note some of the professional and personal qualities which combined clearly to set him apart.

The Development of Urban Climatology

The visually obvious impacts of the growth of cities upon the atmosphere were commented upon and studied from the earliest times but it was not until the invention and development of meteorological instruments such as the thermometer, barometer and anemometer that the more subtle aspects of urban climate could be investigated. Developments from that time can be usefully divided into three periods: first, prior to 1930 simple climatographies of individual weather elements were constructed; second, from 1930 to 1965 description was extended to include greater spatial and temporal variability and linkage of climatic effects to weather and urban structure; third, post-1965 emphasis shifted to seek the physical basis of urban effects and to construct models to simulate them.

PRIOR TO 1930: DISCOVERY AND DESCRIPTION

Luke Howard is credited as the Father of Urban Climatology for his observation, interpretation and reporting of air temperatures within the city of London (at the Royal Society), and at a number of sites, in the then countryside. He reported his findings in two early volumes entitled *The Climate of London* (Howard 1818, 1820) and later in a much expanded second edition of three volumes (Howard 1833). Despite what we would now consider to be unrepresentative exposure of thermometers it is clear that he observed, and correctly recognized, the anomalous warmth of the city (both diurnally and seasonally) that we refer to as the urban heat island. Further he hypothesized concerning the cause of this feature and identified almost all of the mechanisms we now consider responsible. His feat was a most auspicious start to the scientific study of city air.

F. Baer et al. (eds.), Climate in Human Perpsective, 61–75.

The latter half of the nineteenth century saw the rapid expansion of meteorological observing networks around the World and especially in Europe. This provided the essential data for climatological analysis and the establishment of climatographies. A number of the great cities of Europe were studied in this period. These included the thermal climate of Paris by Renou (1855, 1868), Munich by Wittwer (1860), Berlin by Kremser (1886) and a study comparing a number of European cities by Hann (1885). Similar work was begun in North America and Japan in the 1900-1920 period. By the last decade of the nineteenth century, air pollution and its effects on urban fogs and the reduction of sunshine became a central focus especially in England (e.g., F. A. Rusel, 1889; M. J. Rusel, 1892; and Carpenter, 1903). A trend of increasing fog frequency was established and winter fogs and pollution were sufficient to cut urban sunshine duration by one half. Studies of urban effects on other climatological elements followed including the decrease in air pressure by Hann, the change of precipitation by Hellman (1892), and the reduction of both humidity and wind speed first noted by Kremser (1909).

1930 TO 1965: DESCRIPTION AND STATISTICAL LINKAGE

The seeds of change in the field were sown at the turn of the century by T. Homen of Finland and G. Kraus in Bavaria. These were the fathers of micro-climatology concerned with small-scale variability of climates. Their work was built upon by Schmauss and Geiger in Germany and by Schmidt, Lauscher and Steinhauser in Austria. The interest in climates existing within the synoptic station network soon extended to the urban landscape and this was greatly aided by the introduction of the mobile station by Wilhelm Schmidt (1927). This simple technique provided the means to learn a great deal about the details of urban climates and allowed a large number of case studies to be conducted in different cities. The German-Austrian workers of this period also began studies of basic urban climate processes. For example, Lauscher (1934) laid the foundation for an understanding of the effects of building/street geometry on the exchange of long-wave radiation, and Albrecht carried out his very thorough work on street circulation systems.

These developments led to similar work, especially in North America and Japan in the 1930s, but World War II caused a temporary decline in research output (Figure 1). However, the pace more than resumed after the war in response to military sponsorship of research regarding dispersion of gases etc. in urban areas, and because of the need for climatic information to be incorporated in the planning of cities during the reconstruction period, which was also accompanied by rapid industrialization and environmental deterioration. This period produced some of the most detailed urban climatologies of specific cities that are available. Examples include the thermal studies of Bath, U. K. (Balchin and Pye 1947), Uppsala, Sweden (Sundborg 1951), San Francisco (Duckworth and Sandberg 1954), Ogaki City, Japan (Takahashi 1959), Linz, Austria (Lauscher et al. 1959), Vienna (Steinhauser et al. 1959), Kumagaya City, Japan (Kawamura 1964) and London, U. K. (Chandler 1965). These studies are considered 'classics' in the field, moreover some include ideas or observations that are 'ahead of their time'. Sundborg was the first to predict heat island intensity in terms of synoptic weather conditions, an approach also utilized by Duckworth and Sandberg, and Chandler. Even more significantly, Sundborg was the first to couch his ideas on causation of the heat island in an energy budget framework. The Duckworth and Sandberg study is the first full study of the vertical dimension of the heat island, and the works of Takashashi and Kawamura herald attempts to relate features of urban fabric and structure to those of heat island morphology.

Parallel work exploring urban effects on humidity, wind, fog, precipitation and solar radiation continued throughout the period, but the heat island attracted by far the most attention. As a generalization it may be said that the work was characterized by greatly increased description of building layer urban climates and attempts to statistically relate distributions to features of the urban landscape and weather controls.

The field of urban climatology was first reviewed by Kratzer in 1937 and in a much extended version again in 1956. In the same year Landsberg produced his masterful and much-quoted review 'The Climate of Towns'—his first major publication in the field (L1956-a). This was followed in 1962 by his equally significant paper 'City Air - Better or Worse?' which squarely placed the need to understand urban climate in the forum of urban air pollution meteorology (L1962-d).

POST 1965: PHYSICAL PROCESS AND MODELLING

It is not easy to defend the choice of 1965 as a break-point in the development of the subject but there is little doubt that the work in progress at the start of the decade was very different from that at the end. One of the most important reasons for this was the injection of greater interest by meteorologists. This was spurred by increasing concern over the state of the natural environment in general and of urban pollution in particular. The change was further aided by the emergence of a new generation of climatologists who were interested in developing physical climatology and were committed to an energy and water budget framework for their studies. The combination of these ideas and forces led to a range of new foci and an outburst of publications (Figure 2). These included work on basic physical processes in the urban atmosphere such as turbulence, evaporation, cloud microphysics and atmospheric chemistry. Studies also began in an attempt to draw up budgets of energy, mass and momentum for the first time and to include human bioclimate in an integrated scheme. The vertical influence of cities came under full scrutiny using new instruments and observation platforms, and very large-scale experiments were initiated in certain cities. These included the Urban Air Pollution Dynamics Program in New York (Davidson 1967), the Metropolitan Meteorological Experiment (METROMEX) in St. Louis (Changnon et al. 1971) and the Complete Atmospheric Energetics Experiment (CAENEX) in Zaporozhye, USSR (Kondrat'yev 1973).

This period also witnessed the emergence of the first attempts to model urban climates. These include the numerical heat island models of Summers (1964) and Myrup (1969), and the scaled hardware model of Davis and Pearson (1970). Numerical models were also developed for urban wind fields. Taken overall we see a distinct movement towards work emphasizing meteorological process, expansion of vision to include the entire boundary layer, and efforts towards the production of process-response models.

Just as the field as a whole blossomed in this period, so did the contributions of Helmut Landsberg. Many feel that the WMO/WHO Symposium on Urban Climates and Building Climatology held in Brussels in 1968 was a turning-point; it certainly drew many interested workers including Landsberg. He reported the results of a study of the thermal climate of a small courtyard (L1970-e) and summarized principles and knowledge concerning climate and urban planning (L1970-b). Immediately thereafter he produced a series of insightful and comprehensive reviews for professionals (L1970-a,c,d; L1972-b; L1974-d; L1977-c; L1981-a,d) and for more general readers (L1969-c, L197-b). He devoted special attention to the question of applying urban climate information in urban design (L1973-c, L1975-c, L1978-c, L1984-a, L1986-a), and to reporting the results of his long-term observation project on the climate of a growing community (L1972-c,d,e; L1975-b; L1979-a) and the effect of Washington, D.C. on convective precipitation (L1975-d).

These publications represent only a sample of Helmut Landsberg's urban climate writings. He produced more than 35 papers directly on this topic after he was 50 years of age, including his major research monograph *The Urban Climate* (L1981-d) when he was 75, and he contributed seminal work up to his death. The fact that this was but one of a number of fields to which he was devoted is simply staggering.

The Written Contributions of Helmut Landsberg

Everyone has their favorite paper written by Helmut Landsberg. All had a message or two tucked in behind the otherwise obvious title, all were engagingly phrased, all were eminently sensible. No matter how many times he was asked to address the apparently same subject matter he always seemed to have a new angle.

To those meteorologists outside the field of urban climatology he will probably be most known for his authoritative reviews: the classic in *Man's Role in Changing the Face of the Earth* (L1956-a), that in the *Symposium Air Over Cities* (L1962-d), and in *Science* (L1970-c), and in the *World Survey of Climatology* (L1981-a) and his book *The Urban Climate* (L1981-d). All of which displayed his encyclopaedic knowledge and familiarity with work from all over the world and his grasp of the wider context and significance of the work of others. His assessment was always fair and balanced.

To much wider audiences he will be known for his uncanny ability to relate climatology to everyday life in an entertaining and absorbing way without selling the scientific content short. Examples abound including 'Comfortable Living Depends on Microclimate' (L1950-b), 'Study the Microclimate When You Build' (L1950-d), 'Climate and Shelter' (L1977-b) and the widely read popular book *Weather and Health* (L1969-c).

These were the contributions of a dedicated and astute scholar and an enormously talented writer with the common touch.

It is my personal view that Landsberg's most important role within urban climatology, beside the credibility and authority that his association with the field bestowed, was his down-to-earth judgement as to what is relevant and significant. With not a hint of stridency or of pushing his own agenda he provided an unfailing backstop of reason and direction. He was a gentle but firm guiding hand, showing the way by example and encouragement.

The five works of Landsberg that I have chosen to emphasize are examples of this aspect of his influence. I believe they represent some of the works which were closest to his heart: suggestions for future directions in research, the basic needs for monitoring and forecasting, a model of simple experimentation, and principles for incorporating urban climatic knowledge into urban design to benefit the general populace in both the Developed and Developing Worlds.

The first paper, "The Urban Area as a Target for Meteorological Research" (L1974-i), was based in part on an address to the Second Canadian Micrometeorological Conference, and published as part of the Festschrift volume celebrating the 60th birthday of Hermann Flohn. The opening section reads:

> It has been quite natural that questions of a fundamental nature have attracted the curiosity of the scientifically inclined members of the profession and that the unsolved uncertainities of atmospheric menaces to life and property have pre-occupied the practitioners. But some problems have been neglected. Among them are meso- and micrometeorological urban areas. These belong principally to the class of boundary layer studies. The latter have mostly concentrated on the puzzles of turbulence and diffusion but usually have not included forecasts of meso-synoptic systems. Climatic studies of cities and towns have gradually increased in number, scope, and quality. However, on the whole, meteorological research for urban areas has been commensurate neither with the needs nor the economic values involved.
>
> One glance at the population census shows that there has been a worldwide trend toward concentration of people in metropolitan centers. In western countries about 70 percent of the population lives presently in or near towns and cities. Projections to the end of the century predict that 85 percent will live there by the year 2000. This clearly demonstrates that the vast majority of the meteorologist's customers are or will be urban dwellers. We may well ask the question: are we serving these people well?
>
> Let me skirt the answer by first recalling that these masses of individuals have a very high interest in weather. This is demonstrated by the fact that radio and television forecasts have the largest audiences of all programs. Automatic telephone forecast recordings, especially in bad weather, reach extraordinary high call numbers. The fact that climatological information is of great economical value in the design of structures, buildings and houses, their safety, and their heating and air-conditioning plants is gradually penetrating. And a new, insistent customer is the air pollution control officer who is trying to protect the population from acute and chronic health damages caused by high concentration of noxious substances.
>
> Few of these consumers of meteorological services are interested in conditions in a 50 or 100 km radius from their abodes, not even at the local airport, unless they want to travel. But our synoptic forecasts, based on a 100 to 200 km grid, are generally not geared to pinpoint forecasts in space and time to local needs in areas that are generally less than 50 km in diameter. There it is important, and occasionally critical, to know the precise times of precipitation, and amounts, whether in form of rain or snow, the probable concentration of pollutants, temperature and wind values, etc. Our critics and most of us will agree that this is not yet within the state of the art. In making this admission there is, however, no need to take a pessimistic attitude for the future. There is no fundamental reason why such predictions cannot be brought into reach by an adequate program of research.

As I read this paper it seems that more than the urban area was a target! One group in his aim were those meteorologists whose gaze was only fixed at the synoptic or larger scales. Especially in the early 1970's many meteorological services and universities were still bound up with the forecasting missions in aviation and marine weather and the development of numerical weather prediction. Their observations were entirely met by the synoptic and upper air networks together with satellites. Landsberg was pointing out the direct relevance of mesoscale, local and microscale phenomena to the urban public and the need for the profession to address their concerns through greater emphasis on research into urban climate and the development of special forecast services. The message was subtly wrapped up in a delightfully readable educational package.

He was also aiming his guiding hand towards other urban climatologists. Pointing out gaps in knowledge and fruitful avenues for research. Above all he was urging us to keep our sights on more than the esoteric and academic.

The second paper is "Meteorological Observations in Urban Areas" (L1970-d). This paper probably remains to be the essential reading item on this topic. It spells out the variety of observations needed for urban research and engineering. It has been added to by the survey of requirements for air quality networks by Munn (1978) and Vukovich (1978). It was written soon after Landsberg had left the position of Director of the Environmental Data Service of NOAA so he was intimately aware of the extent to which the U.S. network met the urban needs he perceived. The opening paragraph is an especially cogent statement, and I would love to have heard his modern assessment of his final paragraph if he had lived to give the talk he was scheduled to present in celebration of his 80th birthday entitled "2000 AD and Beyond". His final paragraph in the 1970 paper reads as follows:

> Attention to meteorological city networks and their rapid augmentation is as much a meteorological priority task of the outgoing 20th century as the Global Atmospheric Research Program. Projections show that soon 85% of the World's population will reside in urban centers. In some of these the air resources are critically strained already. There is no hopeful sign that the corner has been turned or will soon be turned on pollution control at the source. The people in cities, prone to respiratory ailments, will need reliable meteorological forecasts on the micro-environment as much as long-range global weather outlooks. The meteorological effort needs considerable balancing to achieve that goal.

The third paper, "The Meteorologically Utopian City" (L1973-c), was originally presented as an after-dinner address to those attending the American Meteorological Sociey Conference on Urban Environment in Philadelphia. It is a marvelously spare and unpretentious piece of writing. It is journalistic in its appeal but so very appropriate to the audience he addressed. I remember picking up my copy of the Preprints volume of the Philadelphia conference and remarking to a colleague that it was surprising not to find a contribution by Helmut Landsberg. "Well, he's sixty six now," was the sobering reply, and I acquiesced. Throughout the meeting sessions Landsberg twinkled about, asking a pertinent question here, and pointing out a relevant reference to previous work (usually from other countries) there. The technical sessions were very successful. But for me, and I suspect many others who joined in the standing ovation at the end, Landsberg banquet speech was the binding element to the conference. It gave us our context in more than meteorological terms, it made us proud by association, and it welded us as a community. How thankfully ill-informed my earlier acquiescence proved to be, we were still to be treated to another sixteen papers and a book on urban climate from Landsberg—a successful career for many. Of course that was only 15 percent of his total output in the same 'retirement' period.

Mindful of the impact of this paper I asked Landsberg if he would consider writing a 'tropical Metutopia' paper for the WMO Technical Conference on Urban Climatology and its Applications with Special Regard to Tropical Areas. Again I was concerned that I was asking a seventy-eight year old man to write a challenging paper at short notice and then to travel to Mexico City to participate and present it. Six weeks later I received his paper and of course he was at the center of the conference. The Proceedings appeared only a few weeks after his death and the volume was dedicated to him.

His paper in those Proceedings is titled "Problems of Design for Cities in the Tropics" (L1986-a). It envisions "Swarag - a climatically sensitive tropical town" and is the fourth paper of the five emphasized here.

The application of climatic information to urban and land-use planning was clearly one of Landsberg's fondest wishes. Unfortunately it remained essentially unfulfilled despite his eloquent, even impassioned advocacy. There are many reasons for this (Page 1970; Givoni 1972; Oke 1984). One is the fact that it is most cost effective to institute changes prior to urban development - simple rules to optimize ventilation, solar access, evaporative cooling, pollutant diffusion etc. can then be applied for little or no cost. After development the potential disruption to urban form is usually not acceptable economically and the impacts less easily demonstrated. Landsberg was particularly concerned to get the word through to the developing world where the bulk of urbanization is, and will, be concentrated. Unfortunately it appears that urban growth in many of these cities is not planned at all—it is spontaneous, illegal and desperate. The niceties of climate are hardly on the list of concerns. This did not stop him trying.

In a review of Helmut Landsberg's contribution to urban climatology I could not fail to include his most famous field observational study—the long-term climate changes produced by the growth of Columbia, Maryland. It was his laboratory, providing insights, and illustrative material for many of his writings. Accordingly, the fifth paper emphasized here is "Atmospheric Changes in a Growing Community (the Columbia, Maryland experience)," (L1979-a). It contains the core of the results; for details of method, and additional results the reader should consult the original report (L1975-b). The concluding paragraphs of the 1979 paper follow.

Meteorological factors

The Columbia experiment confirmed in most respects earlier deductions about urban atmospheric alterations. It showed that in an area of small topographic differences there is indeed a measurable difference between conditions before and after urbanization. The rapid development of a heat island can well be taken as the meteorological hallmark of such man-made changes. This seems to be a simple function of size and based on this work, as well as earlier contributions of others (Oke 1973), can be considered as a quantitatively predictable event.

Similarly, the decreases in wind speed and the increased turbulence came as no surprise. However, the urban windfield has far more general variability than the horizontal temperature distribution. One cannot predict with any reliability the local wind alterations that follow town building. Thus, short of windtunnel studies of a city model, one can only give qualitative or generalized estimates of how building activity will influence winds.

Columbia was also no exception to the general development pattern of towns with respect to rapid run-off and increased air pollution. At the present stage of growth there was no discernible influence on precipitation and none can be anticipated for the next decade.

Lessons for planning

The developers of Columbia advertised it as a planned town. It cannot be our province to judge this so far as lay-out, zoning, architecture, and certain amenities are concerned. Only those who live there can evaluate these factors over the long haul. There was also considerable planning to adapt to the physical contours of the landscape. Part of this is reflected in the damming up of low areas to create artificial lakes. These are aesthetic and recreational assets but they are compatible with the summer rainfall pattern in this part of the country only at considerable cost because of the silting problem (a doubtful distinction they share with the Aswan Dam). One can ask if parkland with trees might not have been a more viable alternative.

The lakes were definitely too small to have any appreciable effect on temperatures beyond the immediate shore area. As little as 60 m inland one could not detect any differences. Beyond the steam fog formation, they also had negligible influence on atmospheric moisture parameters.

Obviously, as far as heat island development, wind reduction, and air pollution increases are concerned, Columbia showed no significant advances over other towns of comparable size except that no polluting industries or domestic heating systems were incorporated. But the lack of an adequate, non-polluting public transportation system is likely to haunt this community as it continues to grow. Any town planned for 100,000 inhabitants should incorporate such a system from the beginning.

No imaginative new ideas were notable to cope with parking. Acres of open parking lots in the office center, the shopping mall, the village shopping centers, and the schools are present as in any existing town. If remedies for heat island development are contemplated, such surfaces should be the primary target. Parking garages, above and underground, are one approach.

The heat island is mitigated where daytime shade prevails. Old deciduous trees can contribute greatly. They not only keep the short-wave solar radiation away from the absorbing surfaces but by evapotranspiration use up solar energy. The Columbia area had many old trees but a large number of them succumbed to the bulldozer to make construction easier. Only Symphony Woods and a few isolated trees survived in the area of denser settlement. Construction engineers, architects and landscapers should learn how to preserve as much of the tree growth as possible. Our IR photographs clearly show that every tree stands out as a cool spot in the hot pool of paved and roofed area. As in all new developments home-owners and developers plant new trees to replace natural vegetation, but it takes decades before these grow to a size where they have an influence on the heat balance.

Roofing materials are designed to be water resistant and have a long lifetime. For many climates they should also be reflectant for short-wave radiation. For larger and well supported roofs one can also envisage roof gardens as a remedy for the extreme heat absorption of present heat absorption of present roof types. In so-called primitive cultures sodded roofs were common and had many of the desired merits. They mitigated temperatures in the structures below and certainly did not contribute to a heat island.

The Columbia development also showed that there can be no improvement in air pollution of urbanized areas until transportation by automobiles is either mostly replaced by non-polluting mass transportation or the present type vehicles are radically redesigned or based on non-polluting principles.

In other new towns one might want to look again at principles of land utilization. There are meteorological merits in building up or below ground. Compactness of a town has also advantages. The leg might substitute for the car.

There is no dearth of information on urban micro- and meso- meteorological conditions but this has been largely ignored in planning, although the World Meteorological Organization has for several years in cooperation with other international organizations tried to advance the dialogue among the pertinent professions (1973). If the Columbia experiment has contributed a little to advance toward this goal, it has been worth the effort.

The Special Qualities of Helmut Landsberg

Unfortunately I have no answer to the question as to why Helmut Landsberg was so attracted to urban climatology. My suspicion is that the answer is tied to his obvious enjoyment of, and concern for, people. He demonstrated this also by his interest in the provision of climate services, climate and history, human bioclimatology and human impacts on climate. The city represents a crucible in which these elements interact and find their densest expression.

When one reviews Landsberg's urban climate works it becomes clear that his influence over the field did not derive from his contributions of original or frontier research. There were only three or four such items and none would be considered to have contributed great theoretical or observational findings. The Columbia study is very well known, but mainly for its novel nature rather than for the methodological rigour with which it was designed or the scientific depth of its results. Indeed I suspect that Bill Lowry, who has lovingly noted Landsberg's feet of clay in some of these matters (Lowry 1987), would consider that a golden opportunity was missed by not taking more systematic note of synoptic influences and secular trends in climate. Lowry also mentions examples of Landsberg being "frustratingly less than a giant in matters of mathematical discourse and, as well perhaps, in matters of the comprehension of some of the finer points of theory." This used to bother my small mind because I knew the weight of Landsberg's words. If he made an incorrect statement the misconstructions reverberated around the field for many a year. With some trepidation I pointed out a few such shortcomings in my review of *The Urban Climate* (Oke 1982). I need not have worried; I received a delightful letter from Landsberg thanking me for doing so and chastizing himself.

No, the undoubted eminence of Landsberg stemmed not from his scientific research but from a special combination of intellectual and personal qualities. The former are easy to recognize. His intellect was large and his interests broad and eclectic. The other contributions in this volume and Landsberg's list of publications and services to scientific, governmental and nongovernmental organizations readily attest to this fact. Obviously he received a sound science training in Germany, but you cannot train a Helmut Landsberg. He had an innate curiosity about the nature of the physical world and how humans relate to it. He was an inveterate bibliophile. His command of the literature across many fields was almost incredible. I always knew that before I received the reprints for an article I had published there would be a card from Landsberg requesting a copy, usually with some complimentary remarks. Of particular note was his effort to maintain touch with non-American and foreign language developments. He regularly chided his compatriots for their intellectual chauvinism. His ability to synthesize these disparate and far-flung findings was highly impressive and extremely valuable. One notable example of this was the production of his famous table listing the typical climatic changes caused by a large mid-latitude city. As far as I can discover, it was first published in L1954-a (Table I) and went through many updates until its last appearance in Landsberg's works in L1981-d (Table II). Tracing through the versions published in 1955, 1957, 1962, 1969, and 1981 makes interesting reading, and shows how his estimates altered in line with current thinking and results. For example, the estimate of solar radiation attenuation changes from 10-25% to 0-20% in line with studies showing the alteration of pollution effects as many mid-latitude cities experienced a change from largely particulate and reducing atmospheres to gaseous and oxidizing ones.

These synthetic skills were allied to excellent judgment and an unerring source of common sense. The latter feel for, and knowledge of, the practical everyday needs of people and meteorology's relationship to them was, in my view, Landsberg's true forte. This helped prevent the subject from becoming too esoteric, and his exposition of these values and linkages did inestimable service to the profession by translating research findings and illustrating general principles to a wide audience.

On a purely personal level Landsberg was a true gentleman. A fair, tolerant and moral man of undoubted integrity. Of course, that does not mean that he did not criticize or oppose. He held strong views and upheld his principles—see for example his analysis of the political reasons underlying the forced settlement of nomads by modern governments (L1978-c). He also maintained a steadfast scepticism towards oft-touted panaceas such as the "inevitable but not infallible computer" and the compliance with, and enforcement of, antipollution legislation (for both see L1973-c).

Landsberg exuded a delightful aura. Small of physical stature he was very approachable. He loved people, cherished friends and was the best of company. His good humor and chuckle were infectious, as was his enthusiasm.

Urban climatology has lost its doyen. Above all we shall miss his counsel. He cannot be replaced but his writing leaves us a beacon for the future.

References

Balchin, W. G. V. and N. Pye, 1947: A micro-climatological investigation of Bath and the surrounding district. *Quart. J. Roy. Meteor. Soc.*, **73**, 297-323.

Carpenter, A., 1903: *London fog inquiry*, 1901-02. Report to the Meteor. Council, HMSO, London.

Chandler, T. J., 1965: *The Climate of London*. Hutchison, London, 292 pp.

Changnon, S. A. Jr., Huff, F. A. and R. G. Semonin, 1971: METROMEX: An investigation of inadvertent weather modification. *Bull. Amer. Meteor. Soc.*, **52**, 958-967.

Davidson, B., 1967: A summary of the New York urban air pollution dynamics research program. *J. Air Pollut. Control Assoc.*, **17**, 154-158.

Davis, M. L. and J. E. Pearson, 1970: Modelling urban atmosphere temperature profiles. *Atmos. Environ.*, **4** , 277-288.

Duckworth, F. S. and J. S. Sandberg, 1954: The effect of cities upon horizontal and vertical temperature gradients. *Bull. Amer. Meteor. Soc.*, **35**, 198-207.

Givoni, B., 1972: Urban and building biometeorological research; an overview. *Int. J. Biometeor.*, Suppl. to vol. **16**, 5, Pt. II, 267.

Hann, J. von, 1885: Uber den temperaturunkerscheid zwischen Stadt und Land. *Osterreichische Gesellschaft fur Meteor. Z.*, **20**, 457-462

Hellmann, G., 1892: Resultante des Regenmessversuchfeldes beig Berlin 1885-1891. *Meteor. Z.*, **9**, 173-181.

Howard, L., 1818: *The Climate of London*, Vol.I, London.

Howard, L., 1820: *The Climate of London*, Vol.II, London

Howard, L., 1833: *The Climate of London*, Vols.I-III, Harvey and Dalton, London.

Kawamura, T., 1964: Some consideration on the cause of city temperature at Kumagaya City. *Geog. Rev. Japan*, **37**, 560-565.

Kondrat'yev, K. Ya., 1973: *The Complete Atmospheric Energetics Experiment.* GARP Pub. Ser., No. 12, ICSU, WMO, Geneva.

Kratzer, A., 1937: *Das Stadtklima.* Friedr, Vieweg und Sohn, Braunschweig.

Kratzer, A., 1956: *Das Stadtklima.* Friedr, Vieweg und Sohn, Braunschweig.

Kremser, V., 1886: Vortrag uber das Klima von Berlin. *Festschrift d. 59 Vers. d.d. Naturf. und Artze*, Berlin, 37-38.

Kremser, V., 1909: Ergebnisse vieljahriger Windregistrierungen in Berlin. *Meteor. Z.*, **26**, 259-265.

Lauscher, F., 1934: Warmeausstrahlung uber Horizenteinengung. *Sitz. Akad. in Wien*, **143**, 503-519.

Lauscher, F., Roller, M., Wacha, G., Grammer, M., Weiss, E. and J. W. Frenzel, 1959: Wilterung und Klima von linz. *Welter und Leben, Sonderheft*, **6**, 1-235.

Lowry, W. P. 1987: Helmut Landsberg as urban climatologist and mentor - a loving, personal memoir. Preprint 11th Int. Conf. Biometeor. 34 pp.

Munn, R. E., 1970: Air flow in urban areas. *Urban Climates*, WMO Tech. Note No. 108, 15-39.

Munn, R. E., 1978: *The design of air quality monitoring networks*. Publ. No. EE-7, Institute Environ. Studies, Univ. Toronto, Toronto, 93 pp.

Myrup, L. O., 1969: A numerical model of the urban heat island. *J. Appl. Meteor.*, **8** , 908-918.

Oke, T. R. 1973. City size and the urban heat island. *Atmos. Environ.*, **7**, 769-779.

Oke, T. R., 1982: Book review of The Urban Climate, by H. E. Landsberg, *Bull. Amer. Meteor. Soc.*, **63**, 535-536.

Oke, T. R., 1984: Towards a prescription for the greater use of climatic principles in settlement planning. *Energy and Build.*, **7**, 1-10.

Page, J. K., 1970: The fundamental problems of building climatology considered from the point of view of decision-making by the architect and urban designer. *Building Climatology*, WMO Tech. Note No. 109, 9-21.

Renou, E., 1855: Instructions meteorologiques. *Ann. Soc. Meteor. France*, **3**, 73-160.

Renou, E., 1868: Differences de temperature entre la ville et la campagne. *Ann. Soc. Meteor. France*, 83-97.

Rusel, F. A. R., 1889: Smoke in relation to fogs in London. *Nature*, **39**, 34-36.

Rusel, M. J., 1892: Stadtnebel und ihre Wirkungen. *Meteor. Z.*, **9**, 12-21.

Schmidt, W., 1927: Die verteilung der Minimumtemperaturen in der Frostnacht des 12.5.1927 in Gemeindegebiet von Wien. *Fortschr. Landwirtsch.*, **2**.

Steinhauser, F., Eckel, O., and F. Sauberer, 1959: Klima and Bioklima von Wien. *Wetter and Leben, Sonderheft*, **7** , 1-136.

Summers, P. W., 1964: *An urban ventilation model applied to Montreal.* Unpub. Ph.D. Thesis, Dept. Meteor., McGill Univ. Montreal.

Sundborg, A., 1951: *Climatological studies in Uppsala with special regard to the temperature conditions in the urban area.* Geographica, No. 22, Geog. Instit., Uppsala Univ., Uppsala.

Takahashi, M., 1959: Relation between the air temperature distribution and the density of houses in small cities of Japan. *Geog. Rev. Japan*, **32**, 305-313.

Vukovich, F. M., 1978: Sampling network selection in urban areas. In *Air Pollution Control Part III*, W. Strauss, **ed.**, John Wiley, New York, 387-412.

Wittwer, W. C., 1960: Grundzuge der Klimatologie von Bayern. In *Landes-und Volkskunde d. Kgr. Bayern*, Vol. 1, Munich.

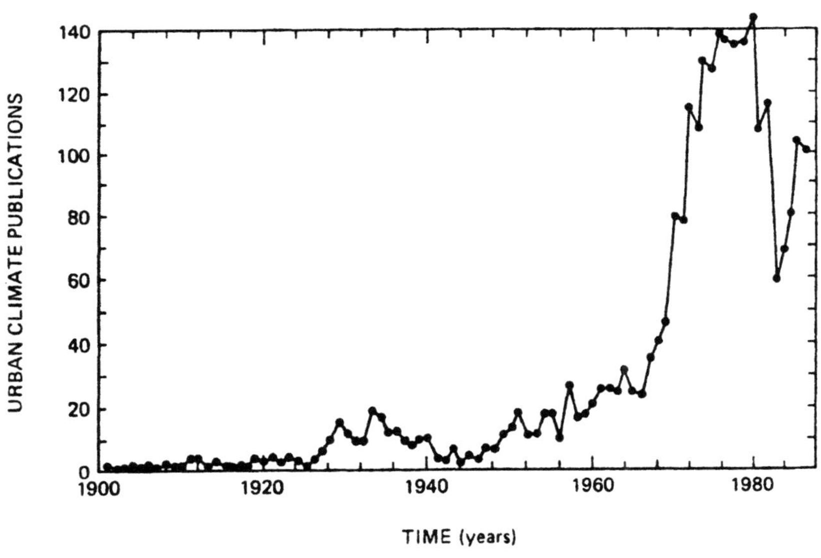

Figure 2 **Trend of total publications in urban climatology (excluding air pollution) in the twentieth century.**

Table I

Climatic Changes Caused by Cities

Element	Change
Temperature	+1 to 2F
Relative humidity	-4 to 8 per cent
Wind	-1 to 2 mi hr $^{-1}$
Sunshine	-100 to 300 hrs yr $^{-1}$
Radiation intensity	-10 to 25 per cent
Illumination	-25 to 50 per cent
Pollution, gaseous (SO_2)	4 to 10 times suburban values
solid, nuclei	15 times country values
solid, dust	10 to 30 times suburban values
organic, bacteria	10 to 12 times country values
Ionization, total	5 to 10 times country values
ratio, heavy/light ions	3 to 9 times country values

Source: L1954-a

Table II Climatic Alterations Produced by Cities

Element	Compared to rural environs
Contaminants:	
condensation nuclei	10 times more
particulates	10 times more
gaseous admixtures	5-25 times more
Radiation:	
total on horizontal surface	0-20% less
ultraviolet, winter	30% less
summer	5-15% less
sunshine duration	5-15% less
Cloudiness:	
clouds	
fog, winter	100% more
summer	30% more
Precipitation:	
amounts	5-15% more
days with >5mm	10% more
snowfall, inner city	5-15% less
lee of city	10% more
thunderstorms	10-15% more
Temperature:	
annual mean	0.5-3.0°C more
winter minima (average)	1-2°C more
summer maxima	1-3°C more
heating degree days	10% less
Relative humidity:	
annual mean	6% less
winter	2% less
summer	8% less
Wind speed:	
annual mean	20-30% less
extreme gusts	10-20% less
calm	5-20% more

Source: L1981-d

BIOMETEOROLOGY

Dennis M. Driscoll

Introduction

I am delighted to participate in this symposium, and to discuss biometeorology, and Helmut Landsberg's contributions to it. His impact upon this branch of applied meteorology was considerable, and what we know about relationships between the atmospheric environment and various organisms, principally man, is due in no small part to his efforts.

My paper is divided into thirds of about equal length. In the first I'll define biometeorology and compartmentalize it, and show how Landsberg's contributions were almost entirely to human biometeorology. In the second part my concern will be with what's being done at present, what the current thrusts are, and who's doing what where. Finally, I'll zero in on Landsberg's specific influences, specify just what his contributions were, and show how human biometeorology as studied and researched today is this way largely because of his influence.

Human Biometeorology

First, some definitions and delineations. One should understand that biometeorology is different from the other kinds of applied meteorology you'll hear about today to the extent that it is highly interdisciplinary. The very word itself suggests a joint effort between the earth scientists—chiefly meteorologists but certainly not restricted to them—and a great variety of scientists representing the bio-side of the pairing. In the kind of biometeorology Landsberg contributed most extensively to, human biometeorology, this means we as meteorologists must work jointly with epidemiologists, clinicians, physicians, physiologists and others. Indeed, I've always envied those of you who work in the middle of meteorology because that's your own "turf" and no one else's.

Biometeorology as a whole can be subdivided according to the organism of interest,—plants, animals, and man. With that distinction we can excise a good portion of biometeorology, the plant and animal parts, because Landsberg's influence and interest were almost exclusively in human biometeorology. I do remember one paper on bird migration early in his career (L1948-a), otherwise his concern was with man as the recipient of various assaults by the atmosphere, some harmful, some beneficial, some palliative, some innocuous. More formally, human biometeorology concerns relationships between natural and artificial geophysical factors and human physiological, pathological, and behavioral responses and adaptations. The atmosphere produces stress, the human responds with strain. We are concerned with those aspects of the atmosphere which are important in promoting stress, as well as the manifestations of strain which are a response to the atmospheric provocations.

F. Baer et al. (eds.), Climate in Human Perpsective, 77–84.

The literature in human biometeorology is spread out over an array of journals and periodicals; everything from the purely medical or physiological journals to those which are primarily meteorological in orientation. At the center of this spectrum in the flagship journal, that of the International Society of Biometeorology.

I've found it convenient to further divide human biometeorology according to four additional aspects. The first of these deals with the extent to which human reaction to the atmospheric environment can be explained by existing knowledge, i.e., by how well the physical transfer mechanism is known. This term (or its German equivalent) appears to have been coined by Konrad Buettner who, along with Landsberg, was one of the pioneers in human biometeorology and also emigrated to this country from Germany. Accordingly, the first of three divisions is quantitative human biometeorology, which describes those relationships for which the physical transfer mechanism is known. Erythema, the reddening of the skin in response to the absorption of solar ultraviolet radiation, is an example, as is the uptake of oxygen in the blood stream, forced as it is by the partial pressure of oxygen in the ambient air. Still another example is sweat rate in response to temperature and humidity excesses. For the most part these cause-effect pairings are quantifiable.

The second of these three divisions is qualitative human biometeorology. This involves human responses which are obviously triggered by atmospheric agents but for which there are no clear-cut explanations, and the physical transfer mechanisms is not well established. Responses are more likely to be chronic than acute. Examples are allergies, particularly those for which the atmosphere acts as a transporting medium. Pollen, dust, and other pathogens are the culprits. Also included are asthma, and the infectious diseases such as influenza and the common cold. Rheumatoid arthritis may also belong in this category because of the nearly universal belief that it is weather-influenced. The lack of a physical transfer mechanism, however, makes this affliction a better candidate for the third category.

In this third category, empirical or statistical human biometeorology, the search is for correlations between the atmosphere and human health and behavior. This is often the first phase of many inquiries, one in which cause and effect are only suggested. One must always be careful not to draw inferences which link these two when the only evidence is circumstantial. Or, as Landsberg himself put it:

> Statistical correlations tell us nothing about causal linkages. They only alert us to the possible existence of a relationship that suggests cause and effect. Much painstaking research needs yet to be done to make the information useful for maintenance of health and for management of diseases (L1976-g).

Both Landsberg and Buettner (1957) insisted that inquiry under this heading must be advanced into the higher categories of qualitative or quantitative biometeorology if progress was to be made. Time and again, as Landsberg surveyed the field, and brought us (including both the scientist and layman) up-to-date, he wrote of things in this order, beginning with the known and proceeding to the unknown, from established to speculative.

The second aspect of human biometeorology - the first, again, was a division of the discipline from established via the physical transfer mechanism to the empirical—

considers whether atmospheric agents responsible for man's reactions act directly or indirectly. At one extreme we note the immediate response to the absorption of solar radiation, or altitude sickness in response to lowered partial pressure of oxygen. At the other extreme, the indirect, are sleeping sickness incurred through the bite of the tsetse fly, whose fecundity is governed by weather conditions. The influenza virus is another example.

The third aspect has to do with the naturalness of the environment. Is this nature's conditions we're dealing with or are they of our own construction? Is it sufficient to utilize the measurements taken in an instrument shelter, or must we specify the conditions in a cryptoclimate such as that in houses, automobiles, factories, offices, or other enclosed places? Aerosols can also be characterized this way, in that the suspended particulates which man puts in the atmosphere can be differentiated from pollens, spores, and other naturally occurring aerosols.

The fourth and final aspect has to do with the consequences of interaction with the atmosphere. Are they good or bad? We usually think of adverse consequences, an approach that can be traced to the writings of Hippocrates. The emphasis of the early Greeks was almost entirely on the negative side. There are, however, both neutral and beneficial aspects of man-environment interaction, although often the latter are only palliative. Neutral interactions include ergonomics, or man in his working environment, clothing requirements (mostly for the military), and sports physiology. Beneficial aspects include spas, retirement colonies, or climatotherapy in general.

Current Research

Current or recent thrusts in human biometeorology is my next concern. What are they? Who's involved, and where are they? First, there is a continuing effort to better characterize the thermal environment of man, especially in particular places such as urban surrounds. These studies involve modeling and/or measuring ambient and radiant temperature, humidity, short- and long-wave radiation, and, less frequently, wind speed or air movement. For the most part these studies are not concerned with physiological or psychological responses or adjustments; the emphasis is on numerical simulations of processes affecting energy balance. The work of Terjung and his associates at UCLA (1971), and of H. Mayer (1978) at the Institut fur Meteorologie der Forstlichen Forschungsanstalit in Munich, are exemplary.

Work also continues toward the development of more suitable biometeorological indexes. These express the combined effects of various atmospheric elements on the thermal comfort of man. Which of these elements are used in an index depends upon whether the interest is in cold or warm discomfort, and the ease with which environmental data can be obtained. At the cold discomfort end only the combined effect of temperature and wind are important for almost all purposes. At the other end the effects of humidity also are of concern, and, in some circumstances, radiant temperature and solar radiation are incorporated. Landsberg provided us with a comprehensive summary of such indexes in his 1972 monograph entitled *The Assessment of Human Bioclimate - a limited review of physical parameters* (L1972-a), at that time the most comprehensive review of the subject. Since then Douglas Lee has topped off a very distinguished career in environmental physiology with his paper entitled "Seventy-five Years of Searching for a Heat Index" (1980).

I can't let a discussion of such indexes go by without some mention of my favorite, wind chill. We've heard so much—probably too much!—about windchill the last several years, and to our detriment, because the application has gotten out of hand. We need to remember that there's a vast difference between the rate at which heat is lost from a plastic can filled with water—as was the case in Paul Siple's original experiment, in frosty, windy, Antarctica—and the rate of heat loss from an appropriately clothed person. The concept has been flagrantly abused by the media. In a sad case of reportorial one-upmanship television weatherpersons (and some major network anchors as well) vie with each other to see who can make the combination of wind and cold seem most devastating. You may remember a National Football League playoff game of a few years back in which San Diego played at Cincinnati. We never were told the temperature, only that the windchill was 39 below! Landsberg and I discussed this subject the last time I saw him, and he agreed that the whole business had gotten out of hand, and that the more recent formulations of windchill and windchill equivalent temperatures by R. G. Steadman (1971) was much better physically grounded, and ought to be used in place of Siple's expressions. They haven't been, of course, because they don't make the cold seem as extreme.

Another biometeorological index suggested recently is that by Larry Kalkstein at the University of Delaware (1987). This normalizes extremes of heat and humidity for each location for which it applies. In effect, he says that sultry conditions in New Orleans can be shrugged off easily, but the same in Chicago can be devastating. An interesting idea, if not carried to extremes.

At the same end of the comfort scale we're still searching for that best of all index. Effective temperature has been rejuvenated. Gagge and his group at the John Pierce Foundation in New Haven (Gonzalez et al. 1974) have developed a measure they call the standard effective temperature. Others have proposed indexes with catchy names: humiture, and humidex. R. G. Steadman, who I mentioned earlier in connection with formulations of windchill, also has developed a sultriness index based on detailed physical considerations (1979). Other papers have reviewed and evaluated these indexes; I've mentioned those of Landsberg and Lee.

A third current or recent thrust in human biometeorology is our continuing efforts to tie morbidity and morality in large metropolitan areas to weather events, often incorporating pollution parameters. The papers of S. H. Schuman (1972), and F. P. Ellis (1978) are exemplary.

Fourth, let me mention that something is finally being done to put man's psychological responses to weather on a firmer footing. This has been the stepchild of biometeorological research, one that investigators seemed to avoid because it's so very difficult - less tractable to analysis and explanation than other areas. It's one thing to count deaths due to a specific cause - a heat wave for example - or measure a chemical constituent of the blood, and quite another to "measure" attitudes toward extremes of heat and cold, or determine how date of birth may affect intelligence, or assess the role of weather in provoking riots. Particularly worthy of mention are studies concerning classroom performance (Auliciems 1976a), the perception of weather (Auliciems 1976b), accidents (Sells and Will 1971), and even the demand for beer (Gander 1972). Here, as is also the case with physiological responses, interpersonal variability makes definitive conclusions all the less likely, a qualification that Landsberg made repeatedly (L1976-g).

Finally, there's some exciting work being done to relate photo period to sleep-wake cycles and reproduction. The experimental work has been done on small rodents, but there appear to be consequences for man; in particular, suggestions have been made for minimizing so-called "jet-lag" for both east- and west-bound airline flights (Reiter 1984).

Landsberg's Contributions

Now I'd like to address Landsberg's specific contributions to human biometeorology. I see him as acting in two ways, of fulfilling two roles. First, he was a provocateur. He pushed and prodded, he pointed the way, and gave shape and direction to the study of man's interaction with his physical environment. He told us what we ought to be doing, and suggested ways of doing it. At the same time, he chronicled events in the field as no one else has. One of the most widely referred to of the books on my shelf is the little paperback called *Weather and Health* (L1969-g). Many times, when someone has asked me how to get a start in human biometeorology, or wanted a quick nontechnical reference, I've said "Here, read this."

And there were of course more scholarly contributions. The first evidence I can find of his interest in the subject was a 1935 paper dealing with air quality in rooms (L1935-e). A short time later he wrote the first of what were to be two papers on, of all things, bed climate (L1938-f). Not so surprising, really, because it was always Helmut's contention that we should study the real environment of man, the cryptoclimate, and not that measured by instruments in a shelter five to six feet off the ground.

As early as 1939 Landsberg suggested that the elements the (then) Weather Bureau measures may not be those most important to biometeorology (L1939-d). He was among the first to suggest that electrical properties of the atmosphere may have something to do with man's health and well-being (L1941-b). To quote, "It is surprising to contemplate that the nation that could claim Benjamin Franklin, the discoverer of atmospheric space charges, as the first internationally recognized scientist, has so neglected atmospheric electricity in the lower atmosphere as a subject for research." Studies pioneered by the Germans in the 50's on sferics and air ionization, and continuing virtually world-wide, attest to his prescience. The study of cryptoclimates, those of enclosed spaces such as houses, automobiles, and work places, was also a concern of Landsberg (L1964-a).

Landsberg continually stressed the need for better indexes of man-environment relations. He applauded the work of R. G. Steadman (1971, 1979) and others who have attempted to develop indexes, at both ends of the comfort spectrum, which are better grounded physically. At the same time he emphasized the need for experimental verification of these indexes, a plea that for the most part has gone unheeded. He pointed to then neglected areas of research such as skin cancer and its geographical variation (L1960-a). Subsequently we learned of the rather close relationship between the incidence of malignant melanoma and latitude in the United States (Brinckerhoff 1966). He advanced the idea that the time rate of change of atmospheric elements may be as important as absolute values, an assertion that is corroborated when we examine mortality and morbidity data from large metropolitan areas and find that they increase at times of changing weather (L1968-b). He asked if the spatial distribution of the climatic elements has any causal relationship to the distribution of the races of man: an excursion into anthropological biometeorology (personal communication).

As I noted in an earlier section, Landsberg cautioned against unwarranted conclusions based on statistical methods alone. In this connection he acted as a liaison between investigators for whom weather influences are virtually taken for granted, the European —especially German—biometeorological community, and the more skeptical investigators on this side of the Atlantic. To the former, I suspect he acted like the proverbial man from Missouri: "Show me," he said, "demonstrate that these associations are beyond chance occurrence." To us he suggested that there might be something here, something worth pursuing in these European studies which showed, time and again, a relationship between seemingly innocuous weather and a variety of pathological and psychological reactions. Let's see if we can find the same responses in this country, is the charge he gave us.

I must tell you a little story here that I found humorous, although it's a bit of a digression. For you to understand this I need to amplify a statement I made a moment ago. There is this transatlantic dichotomy in belief in weather influences. In Europe, particularly in eastern Europe, weather influences appear to be taken for granted. The emphasis on research there has been to document the extent and nature of this relationship, rather than determine its validity. Furthermore, the farther one goes eastward in Europe the stronger the belief. Indeed, in Czechoslovakia, Poland, and Yugoslavia, weather influences seem to be taken for granted, even to the extent that medical-meteorological forecasting is practiced, and physicians schedule surgery only after they've consulted the latest weather report. We tend, in this country, to be doubting Thomases, and much more mechanical in our approach to etiology. As I've noted, Landsberg and I have tried to advance the idea that there might just be something here that needs closer examination. You can imagine my delight, then, when I read the headline on a very short story in my local paper: "Surgery Scheduled Around Cold Fronts." The dateline was Apalachicola, Florida, and the first line gave me hope that our efforts were not in vain: "A vital step in preparation time before an operation at Weems Memorial Hospital is the weather report." My reaction turned abruptly to disappointment as I read the next line: "A renovation project has left the operating room with a huge hole two feet wide and fifty feet long in the roof."

Landsberg was of course the recipient of a number of honors and awards during his very productive lifetime. I want to emphasize those given him in connection with his contributions to biometeorology. He was president of the American Institute of Medical Climatology for many years, and was instrumental in its founding. He was chairman of the Publications Committee of the International Society of Biometeorology (ISB), and also helped found that organization. He was awarded the William F. Petersen Foundation Gold Medal in 1983 for his many contributions to the field of human and urban biometeorology, and for his great assistance in the development of the ISB. The most prestigious award was that given to him by the American Meteorological Society. He was the recipient in 1964 of the Award for Outstanding Achievement in Bioclimatology. That citation reads as follows:

The Award for Outstanding Achievement in Bioclimatology is given to Helmut E. Landsberg for his contributions to the science of bioclimatology and for his vigorous and effective efforts to bring attention to this field in proportion to its scientific and public importance. Dr. Landsberg not only has climate of the biosphere, but also has brought out clearly the main biological, ecological and medical factors of the micro-climate that affect the health, happiness, efficiency and safety of the human organism. In view of the fact that man's knowledge should be applied primarily to man in his habitat we wish to recognize our appreciation of Dr. Landsberg's efforts to increase our understanding of the normal human environment and its effect on our lives and activities.

References

Auliciems, A., 1976: Classroom performance as a function of thermal comfort. *Int. J. Biometeor.*, **16**(3), 233-246.

Auliciems, A., 1976: Weather perception, a subtropical winter study. *Weather*, **31**(9), 312-316.

Brinckerhoff, G. D., 1966: *A note on solar radiation and skin cancer deaths in the United States*. Environ. Data Service, Tech. Note 33-EDS-1, U. S. Department of Commerce.

Buettner, K. J. K., 1957: Aim and purpose of symposium on bioclimatology. *Proc. Natl. Acad. Sci.,* **16**, 607-608.

Ellis, F. P, 1978: Mortality in the elderly in a heat wave in New York, August 1975. *Environ. Res.*, **15**(3), 504-512.

Gander, R. S., 1972: Demand for beer as a function of weather in the British Isles. *Weather forecasting for agriculture and industry, a symposium.*, J. A. Taylor, Ed., Associated University Presses, Cranbury, New Jersey, 184-194.

Gonzalez, R. R., Y. Nishi and A. P. Gagge, 1974: Experimental evaluation of standard effective temperature; a new biometeorological index of man's thermal discomfort. *Int. J. Biometeor.*, **18**(1), 1-5.

Kalkstein, L. S. and K. Valimont, 1987: An evaluation of winter weather severity in the United States using the weather stress index. *Bull. Amer. Meteor. Soc.*, **68**(12), 1535-1540.

Lee, D. H. K., 1980: Seventy-five years of searching for a heat index. *Environ. Res.*, **22**, 331-356.

Mayer, H., 1978: Human heat balance in the summer in a forest atmosphere, a city atmosphere, and a seacoast atmosphere. *Arch. Meteor. Geophys. Bioklim.*, **25**(2), 177-189.

Reiter, R. J., 1984: Impact of photoperiodic information on pineal metabolism and physiology in rodents. *Abstracts 10th Int. Congress Biometeor.*, Tokyo.

Schuman, S. H., 1972: Patterns of urban heat deaths and implications for prevention: Data from New York and St. Louis during July 1966. *Environ. Res.*, **5**(1), 85-92.

Sells, S. B., and D. P. Will, 1971: Accidents, police incidents, and weather, a further study of the city of Fort Worth, Texas, 1968. Tech. Report No. 15, Inst. Behavioral Res., Texas Christian Univ.

Steadman, R. G., 1971: Indices of windchill of clothed persons. *J. Appl. Meteor.*, **10**, 674-683.

Steadman, R. G., 1979: The assessment of sultriness, Part I: A temperature-humidity-index based on human physiology and clothing science; Part II: Effects of wind, extra radiation and barometric pressure on apparent temperature. *J. Appl. Meteor.*, **18**(7), 861-873, 874-885.

Terjung, L. S., 1971: Potential solar radiation climates of man. *Ann. Assoc. Amer. Geogr.*, **61**(3), 481-500.

CLIMATE SERVICES

William H. Haggard

Introduction

Helmut Landsberg was the architect of the Nation's modern climate services. After many years of research, publication, and leadership in diverse endeavors relating to climatology, he was invited—in 1954—to modernize the U. S. Weather Bureau's climatological service.

Landsberg demonstrated an impressive ability to organize, staff and operate the Climatological Services Division, which later (1956) became the Office of Climatology. He established an aggressive field program, backed up by modernized data processing facilities and a meaningful publications program. These soon gained wide recognition for one of the most modern and responsive climatological services in the world.

Landsberg's office at Suitland, Maryland, became the focal point for staff selection, program design, policy making, and implementation of the new climatological program. He built an organization consisting of an effective mix of staff and operational people given authority, responsibility and guidance to make the program work and the policies effective. His quiet competence in team building soon paid off and effective transitions from the old to the new functions and procedures took place.

I personally was very overjoyed when in 1954 Landsberg saw fit to hire me for a place on the transition team. At that time he gave me one small piece of philosophy. He said, "Always have a plan and whenever an opportunity presents, put a little piece of that plan into effect and sooner or later you will have your whole plan in place." And that's a philosophy he practiced and a philosophy that worked very well in the climatological services. Landsberg was not only an academician, an author, but a very pragmatic and practical person who wanted the data to be used and to be used wisely, and he knew that it couldn't be unless the data were available to the user in the best possible form.

Field Services

On June 15, 1954, Landsberg submitted a seven-page modernization plan to Dr. Francis W. Reichelderfer, the Chief of the U.S. Weather Bureau.

One part of the promptly-approved plan addressed how to make information in the field available to users in the field. Four additional Area Climatologist positions were established with emphasis on indicated problem areas:

- Northwest Area (forecast problems)
- Southwest Area (arid region problems)
- Southeast Area (marine problems)
- Northeast Area (industrial problems)

F. Baer et al. (eds.), Climate in Human Perpsective, 85–90.

The key item in the field services part of the plan was the appointment of State and Territorial Climatologists to serve all states and major territories. As you all know, in a 1973 budget crunch, that program was simply cut off, and it has slowly come back, under auspices of state agencies and universities. There is now an association, the American Association of State Climatologists. Service effectiveness is not as consistent as it would be if the federal funds had stayed with the program, but the vision that Landsberg had for these focal points in the field to make the data useful and available to the users has persisted—the real need was there. Other field service principles set forth in the plan included:

- Locate Area and State Climatologists where there is maximum opportunity for collaboration with state officials, state universities, agricultural experiment stations, etc.
- Fill positions of Area and State Climatologists only with personnel having climatological background and experience.
- Make climatological work the only function of State and Area Climatologists.
- Set positions of Area and State Climatologists at a good civil service grade.

I hear he had a little problem with that last principle, but the program proceeded.

Data Processing

How many of you remember FOSDIC? I don't mean Fearless Fosdic in the comic strips. I mean Film Optical Sensing Device for Input to Computers. You may know that after the New Orleans Tabulating Unit was moved to Asheville, North Carolina and became the National Weather Records Center, accumulations of punched cards steadily increased at Asheville until the building literally began to collapse under the weight of half a billion punch cards. The very serious question of what to do about miniaturizing these data came into being. And as an interim step, recognizing that new technologies were rapidly evolving, the Film Optical Sensing Device for Input to Computers became not only part of the U.S. Bureau of the Census, but part of the program at the National Weather Records Center. There was a picture published in 1960, showing Landsberg looking at a 100-foot roll of film containing thousands of punch card images which were read on the then advanced - now totally obsolete - electronic equipment in the background.

Throughout the growth and implementation of improved and expanded climate services, Landsberg constantly sought and implemented more efficient and productive procedures and organizational structures to provide climatic services. One landmark example was the consolidation of all climatological records processing activities at the National Weather Records Center (NWRC) in Asheville, North Carolina, from three Weather Records Processing Centers located at Chattanooga, Tennessee; Kansas City, Missouri; and San Francisco, California. A memorandum dated April 25, 1962 correctly anticipated the advantages to be gained by this change as follows:

(1) Utilization of higher order machines will results in greater efficiency. The ever-increasing number of stations requires greater processing potential. It is only through greater efficiency that the additional work load can be accomplished without increase in staff.

(2) Use of the modern machines will make possible a new level of operation from which the Bureau can proceed to develop more and more useful by-products from its routine processing.

(3) Location of data processing in the proximity of the NWRC printing plant will facilitate the earliest possible release of published data with a minimum of difficulty because all problems related to the photographic copy can be resolved locally.

(4) Consolidation will eliminate the difficulties associated with the somewhat different processing procedures in practice at the three currently operating units. This developed from the fact that work loads are different and each unit was forced to have at least one of each type of the basic machines. Thus it became more efficient for a given processing center to follow a procedure utilizing its least-loaded piece of equipment rather than to use a machine simply because it was done that way at an ideal center.

(5) Consolidation at one location will make current records more readily available for emergency use. At the present time it is not uncommon to call for records at a time when it is impossible to be certain of their location; they are upon many occasions in transit to the NWRC after basic processing at one of the local centers.

(6) Moving the entire program to the NWRC will practically eliminate the difficulties of the peak annual load now experienced at the present centers. The NWRC simply has so much more potential that such extra work can become just one more non-routine job.

(7) Administration will be simplified. Instead of having the work administered through three or more Regional Administrative Offices the entire administration will take place at the NWRC under the direction of the Office of Climatology.

(8) Other administrative functions (management of the station networks that have naturally descended upon the present centers) would be returned to the Regional Administrative Offices where they belong and where they can properly be controlled by the Observations and Station Facilities Division.

Publications

A third area in which Landsberg constantly sought better ways to improve climatic services was through improved publications.

The content of existing subscription publications was improved, including monthly and annual *Climatological Data* for each state; *Local Climatological Data* (monthly and annual) for nearly 300 cities; *Hourly Precipitation Data* (monthly and annual) for selected stations by state; *Storm Data*, a National monthly publication of unusual weather events; the *Weekly Weather and Crop Bulletin*, a joint Commerce/Agriculture publication; and *Daily Weather Maps, Weekly Series*.

New decennial and period publications were started, including *Comparative Climatic Data*, means and extremes of climatic data for 284 cities; decadely updated normals; *Climatography of the United States*, monthly and annual mean data and sequential data for 1800 cities; *Climates of the States*, for all states and Puerto Rico; and *Airport Climatological Summaries*, for 163 airports across the U.S.

New historical publications were compiled such as *Climates of the World*, with narrative summaries, maps and statistical data for selected world locations; *Climatic Atlas of the United States*, with monthly and annual U.S. Climatic maps and tables; *Climatic Summaries for Ocean Data Buoys*; *High Altitude Meteorological Data*; *Marine Climatological Summaries*; *Monthly Solar Radiation Data*; *Climatological and Oceanographic Atlas for Mariners* published jointly with the U.S. Navy Hydrographic Office; and a host of others.

An aspect that many of you may or may not be aware of is that Landsberg took advantage of opportunities, be they great things or disastrous things that occurred. You may recall that in 1956, or thereabouts, two ships were on a collision course off Nantucket. The *Stockholm* and *Andrea Doria* collided. It was a terrible marine disaster and was much in the news. Incidentally, there's a fairly recent book entitled, *Collision Course* which is the story of the collision and its aftermaths. Lawyers got into the act and everybody was looking for data. The resources of the National Climatic Data Center, which was then the National Weather Records Center, were sorely taxed to provide xerox copies of the marine data that were available and might have pertained to the accident. And Landsberg had an idea, and out of that idea came a publication that just celebrated its 25th anniversary, *The Mariners Weather Log*. He said there's such a vital interest in good marine climatic data being made available to the user community, that we have to have some kind of a publication that does this on a regular basis, so we don't have to strain the resources of the government to produce data when we have a disaster of this type—and so that the user doesn't have to go searching as hard for the data. So regular features in the *Mariners Weather Log* include the local climatological data of the ocean data buoys, gale tables from ships' weather observations, and descriptions of the more important weather events of the oceans. A monthly rough log is followed later by a smooth log describing the weather events. This publication has survived over 25 years even though it's a free publication and free publications in the government have a hard time surviving. It has survived because it serves a user community that is important and that continually responds to the essential questions coming out to them, "Do you need this publication and why do you need it?" Because the people who use it and need it do respond, the publication has persisted and it is one of Landsberg's many ways of getting data to the users by implementing a small piece of a total plan.

Conclusion

Throughout his tenure as U.S. Weather Bureau Director of Climatology, Landsberg remained true to his maxim of always having a plan ready. An especially noteworthy planning document was a report to the U.S. Interdepartmental Committee on Atmospheric Sciences dated August 15, 1961, and entitled *An Outline for Climatology*, 1961-1970. This 35-page document contained recommendations for seven major avenues of effort, described the status of each, enumerated the tools needed, and spelled out specific tasks to be accomplished relating to:

- Water Resources and Droughts,
- Biological Influences of the Atmospheric Environment,
- Synoptic and Dynamic Climatology,
- Statistical Climatology,
- Climatic Trends,
- Micro and Mesoclimatology, and
- General Climatology.

The personnel and funding requirements for each task were specified. The contributions of individuals and groups were acknowledged.

A condensation of the report was published in April 1962 (L1962-f). Landsberg's summary discussion on climatic trends follows.

> Trends are the greatest puzzle in climatology. Why and how have climates changed? Procedures and the type of answers we can hope to obtain depend greatly on the time interval with which we are concerned.
>
> Closest to us are the trends and fluctuations during the interval of instrumental observations. We hope to find, through continued analysis of long-term records, clues to the causes for fluctuations of varying time-length. As yet, no decisions can be reached as to whether the causes are terrestrial or extra-terrestrial, or a combination of these. In each case a multiplicity of causes has been suspected, but no firm answers are at hand. Once causes have been found—whether they are terrestrial tele-connections or backfeed mechanisms, or diastrophic or catastrophic events, or are embodied in celestial mechanics and solar phenomena—a hope for longer-range predictions can be held.
>
> Recognition of true trends would be of considerable impact on planning, especially for the water supply problem.
>
> Obviously, there is also a crying need to find what human interference has done to natural climatic conditions, be it through urbanization, cultivation, or modification of the atmospheric composition. We should at least learn how much we are tampering with our destiny.
>
> The contemporary fluctuations or those of the recent past might enable us to "calibrate" indirect measures, such as those employed in dendrochronology and palynology, varve studies, and interpretation of animal associations in climatic terms. This requires an interdisciplinary effort which will have important bearing particularly on the problems of the post-Pleistocene climates. Some of it might even carry further back into paleoclimatology.
>
> In this last field a truly heroic effort is involved. Interpretation of rocks and fossils in terms of climate is among the most complex scientific tasks. Many side-issues are

involved, not the least among them being pole positions, relative position of continents, and extend of land and oceans. The most precise data are likely to come from isotope investigations of the type so successfully carried out by Emiliani in 1957. Much further work in this direction is warranted.

Even if it may prove very difficult to unravel the climatic mysteries of the past, it is imperative that all possible assistance be provided to the climatological historian of the future. This requires meticulous documentation of current observations—and as far back as records will permit. Also, station stability for at least a restricted network is absolutely imperative. These stations should be essentially complete observatories, providing continuous records of all atmospheric variables. The establishment of additional stations in nature reservations should be promoted. There phenological records can be kept in parallel with the climatological data. Although the benefits of such a policy will be decades away, a beginning has to be made *now*. The groundwork for some tasks in science must be laid on a long-term basis, and the study of climatic trends is one of them.

Landsberg's dozen years as the prime mover of improved climate services resulted in solid, basic, efficient and effective programs which have endured to this date. Our nation owes Landsberg a debt of gratitude for his foresight, integrity and leadership which established these basic, vital and ongoing climatic services.

THE CLIMYTHOLOGY OF AMERICA[1]

David M. Ludlum

History is full of myths, and so is climatology. Every generation of historians gives rise to a revisionist school that reinterprets the past in light of new material and facts. Sometimes the revisions join the body of history; other times they are revised by the next generation. Overall, the process leads to a richer and more truthful history.

The settlement of America produced a series of myths about the climate of different regions of our country. Even before the first British settlements in North America, Europeans held certain concepts concerning the supposed climate of the New World, and those concepts greatly influenced their efforts to establish colonies from Newfoundland to the Carolinas.

Once the seaboard was occupied, new myths arose about the lands west of the Allegheny Mountains. Other unfounded beliefs appeared to influence the occupation of the Mississippi Valley and Great Plains until, in the last decade of the nineteenth century, the land office in Washington officially declared the frontier closed, though much territory remained unsettled. Most of this however, was thought to be wasteland unsuitable for cultivation. This belief would be dispelled in the next century by the introduction of scientific methods of agriculture and the construction of huge irrigation projects.

The Equal-Latitude Myth

The intellectual content of climatology had made little progress from the time of Ptolemy, the Greek astronomer and geographer of the second century A.D., to the year 1601, which marked the beginning of the century of colonization of North America by the English and the French. The concept of clima, or parallel bands around the world which shared comparable temperatures and hence weather conditions, was the generally accepted view of global arrangements. So much so, in fact, that the word *clima* was used by English writers interchangeably with "latitude." This gave rise to what I shall call the equal-latitude myth. The planners and backers of the new colonies held to the classical view of the distribution of global temperatures and thus were greatly surprised and chagrined when their environmental expectations were not met by the realities of the New World. The French were perplexed by the harsh winter conditions they met in Nova Scotia and the St. Lawrence Valley because both lay at the same latitudes as northern and central France. The British ultimately gave up constant efforts to settle Newfoundland in the early years of the seventeenth century because of the severe winters, despite the fact that it lay at the same latitude as southernmost England, where winters were usually moderate in temperature.

[1]Advance publication in *Weatherwise*, **40**(5) October 1987, 255-259

F. Baer et al. (eds.), Climate in Human Perpsective, 91–96.

The history of all the British colonies from Maine to the Carolinas ran much the same. The commercial backers of each colony expressed surprise and dismay that these settlements, though at the latitudes of France and Spain, could not produce the exotic agricultural products of those countries. Believing Virginia to have a Mediterranean climate, the proprietors tried silk culture until the realities of the winter killed all hopes of producing such a tropical product.

Almost a century passed before the backers of the colonies realized that the American climate differed from the European at the same latitudes. By the beginning of the eighteenth century, a more realistic viewpoint prevailed about the climate of the New World. Facts replaced the equal-latitude myth.

The Climate Change Myth

During the first two centuries of settlement of the American seaboard, a popular misconception arose about the observed climate. Where were the record snows of yesteryear? Why did we not have the harsh winters so often mentioned by grandfather and great-grandfather? Many homespun philosophers pondered these questions and suggested answers. Though no actual facts were brought forth, most colonists believed that conditions had grown milder and that the seasons had changed, with spring coming later and autumn lasting longer.

These ideas were expressed in an article by Dr. Hugh Williamson of North Carolina in the first issue of the *Transactions of the American Philosophical Society* in 1771: "An attempt to account for the change observed in the Middle Colonies in North America."

Williamson's thesis was that the cutting down of the forests for farms and settlements had produced a warming of the soil for two reasons. First, the felling of the trees allowed easterly winds to penetrate more deeply into the country, bringing temperate marine influences inland. Second, the bare soil received and stored more solar heat than did forested lands, an snow melted more quickly when exposed to direct sunlight.

In addition, some colonials suggested that the rise of urban communities with heated buildings and smokepots was leading to a milder climate, as they claimed had occurred in Europe. These ideas were the first of many about climate change that were to rise and claim a body of believers among Americans.

The Ohio Country Myth

After almost 200 years of English settlement along the Atlantic seaboard, the vast interior of the North American continent remained a *terra incognita* as far as an exact knowledge of its geography and climate was concerned. The French had sent voyagers, couriers de bois and missionaries deep into the interior, but their first-hand knowledge of the conditions encountered did not reach the seaboard-bound British. Though the barrier of the Appalachian Mountains was breached during the war years that marked the closing decades of the eighteenth century, few scientific men went westward to observe and report on the physical and atmospheric geography of the interior.

A vigorous controversy as to the nature of the climate of the Ohio Country beyond the Allegheny Mountains arose as the century drew to a close and continued to spark lively arguments well into the next century. The controversy became known as the Ohio Country myth.

Between October 1795 and June 1796, Constantin François de Chasseboeuf, Comte de Volney, traveled from Washington, D.C. to Vincennes on the Wabash River in Indiana. He was familiar with Thomas Jefferson's view, expressed in his *Notes on the State of Virginia*, that the annual temperature west of the mountains was several degrees warmer than at the same latitude east of the mountains along the Atlantic seaboard. Jefferson based his opinion on the different types of plants thriving on opposite sides of the mountains. Volney's seeming confirmation of Jefferson's opinion received wide dissemination in the *View of the Climate and Soil of the United States*, published in London and Paris in 1804. The first refutation of the ideas promulgated by Volney came from Dr. Daniel Drake in *Notices concerning Cincinnati*, published in 1810, which produced actual comparative temperature readings. Others soon took up their scientific cudgels. In an address before the Albany Institute in 1823, Dr. Lewis Beck took each of Volney's statements and demolished them with facts from more recent material.

William Darby, in his *View of the United States: Historical, Geographical and Statistical* (1828), referred to Volney's "by no means innoxious vulgar error." As late as 1842, Dr. Samuel Forry, in the first climatological survey to employ meteorological observations, felt constrained to criticize Volney's opinions as being "barren of precise data."

In 1857, Lorin Blodget put the Ohio Country myth to final rest in his comprehensive *Climatology of the United States*: "The early distinction between the Atlantic States and the Mississippi has been quite dropped, as the progress of observation has shown them to be essentially the same, or to differ only in unimportant particulars."

The Great American Desert Myth

"When I was a schoolboy my map of the United States showed between the Missouri River and the Rocky Mountains a long, broad white blotch, upon which was printed in small capitals 'THE GREAT AMERICAN DESERT - UNEXPLORED.'" So wrote Colonel Richard Irving Dodge in 1877 when commencing his revealing survey, *The Great Plains of the Great West*. He concluded: "What was then 'unexplored' is now almost thoroughly known. What was regarded as a desert supports, in some portions, thriving populations. The blotch of thirty years ago is now known as 'The Plains'."

Sergeant John Ordway, who had accompanied Lewis and Clark in 1804, has stated "... this country may with propriety be called the Deserts of North America." Captain Zebulon Pike in exploring the headwaters of the Arkansas River had declared that "... these vast plains of the western hemisphere may become in time celebrated as the sandy deserts of Africa." And Major Stephen H. Long had written, "... the Great Desert at the Base of the Rocky Mountains ... is almost wholly unfit for cultivation, and of course uninhabitable...."

When Lorin Blodget published his comprehensive *Climatology of the United States* in 1857, he marked a zone running east of the 100 W meridian on his precipitation chart "the eastern limit of the dry plains," and labeled the area of western Kansas and Nebraska "the Desert Plains."

Following the Civil War, a counterattack was launched on the pessimistic opinion about the future of the plains. The pressure for new lands to settle caused a change of view regarding the farming possibilities of the plains west of the Missouri River. Optimistic projections were penned by enthusiastic travelers, booster-type editors and eager business promoters. Their hopes were bolstered by several years of above-normal rainfall in the late 1860s and early 1870s. The concept that "Rain Follows the Plough" was broadcast in chamber-of-commerce style by agricultural improvement societies and business enterprises. This was the "Garden Myth" - that planting trees and crops on the dry plains would result in increased rainfall in a self-perpetuating manner. The climate pendulum, however, underwent several swings from adequate to inadequate rainfall until a nadir was reached in the late 1880s and early 1890s, resulting in disaster for the many cattle ranchers and the abandonment of farming in much of western Kansas and western Nebraska.

The occupation of the central plains by farmers, the western plains by cattlemen, the mountains by miners and the Pacific Northwest by lumbermen brought more adequate knowledge of the actual climates of these regions. The filling in of the nation's climatological charts was completed about 1890, when the availability of free land ended and the frontier was considered closed.

The Southern California Health Myth

During the first 30 years of American settlement Southern California remained a frontier country with ranching and agriculture dominating the economy. The last two decades of the century, however, brought a change. Promoters and developers exploited the region's prime natural attraction, a beneficent climate, to make it the health frontier of the United States. Its favorable features were widely promoted in a tidal wave of publicity, and hordes of Easterners responded by migrating to the promised land in search of restored health. Thanks to man's ingenuity, the barren outlands had suddenly become habitable and even attractive.

During the decades from 1850 to 1880, native Angelenos might have been forgiven for doubting their climate would turn out to be the most promising feature of the region. Damaging floods occurred in 1862 and 1868, devastating droughts came in 1862-64 and 1876-77 and a long spell of recurrent cold weather in the late 1870s and early 1880s set many still-standing date records for coldness. In addition, a destructive earthquake struck in 1857 and every year there were "tremblos."

Despite the lack of knowledge of the effect of California's climate on disease, publicity for the region's salubrity soon poured forth. A pamphlet entitled, *Southern California: The Italy of America*, claimed for the area the "only perfect climate in the world and the grandest scenery under the sun." The *Los Angeles Star* in 1872 carried an article, "Land of Glorious Sunsets," which was considered by historian Oscar O. Winther (in 1946) as "the opening trumpet blast of a climate promotion campaign that has not ended."

Concerted efforts to attract visitors and settlers became an increasingly active industry in the 1880s. The Local Chamber of Commerce was careful to point out that not all parts of California enjoyed the salubrious climate claimed for the southern region. The results soon became apparent. A great boom in real estate and business developed in the mid-1880s, similar to those previously experienced in other sections of the western frontier country.

In the 1890s, climate continued to be the principal pitch of promotion agencies. In 1892, the Southern California Information Bureau asserted: "... we sell the climate at so much an acre and throw in the land." To a complaint that the region had nothing to sell except climate, one enthusiast declared: "That's right, and we sell it, too - $10 for an acre of land, $490 an acre for the climate."

The health angle and longevity prospects were emphasized in the promotional publications of the 1890s. Dr. Peter C. Remondino stated the extreme claim for the region in his book, *The Mediterranean Shores of America: Southern California*: "from my personal observations, I can say that at least an extra ten years' lease on life is gained by a removal to this coast from the Eastern States; not ten years to be added with its extra weight of age and infirmity, but ten years more with additional benefit of feeling ten years younger during the time."

They came at first by the thousands, and finally by the millions; today more than 15 million people live in Southern California where a century ago there were only 32,000.

Ironically, the concentration of population with attendant urban sprawl and congested freeways affected the climate in a way none of its promoters of the late 1800s foresaw. The effusions of millions of combustion engines, trapped in the area's natural basins by the almost daily inversions in the lower atmosphere, have created smog conditions detrimental to health.

Alaskan Climythology

The bill for $7,200,000 to pay for Alaska "loosed a storm in the House of Representatives. I shall not attempt to say whether it was a hurricane or tornado, but it was accompanied by a lot of wind, by a great flood — a flood of oratory and some verbal thunder," declared Senator Ernest Gruening at a meeting of the American Meteorological Society at the University of Alaska on June 27, 1962. The former Russian colony was portrayed as a "frozen waste with a savage climate, where little or nothing could grow, and where few could or would live."

Typical of the statements of these pioneer climythologists was that of Benjamin F. Loan of St. Louis, who declared: "... the acquisition of this inhospitable and barren waste will never add a dollar to the wealth of our country or furnish any homes to our people. It is utterly worthless. ... To suppose that anyone would leave the United States...to seek a home...in the regions of perpetual snow is simply to suppose such a person insane."

Another climatic pessimist, Representative Orange Ferris of Glens Falls, New York, asserted that Alaska "is a barren and unproductive region covered with ice and snow" and "will never be populated by an enterprising people."

A representative from New York, Dennis McCarthy of Syracuse, cited "reports that every foot of the soil of Alaska is frozen from five to six feet in depth" and ventured that his colleagues would soon hear that Greenland was on the market.

And the minority report of the House Committee on Foreign Relations, in a scathing denunciation, declared Alaska "had no capacity as an agricultural country ... no value as a mineral country ... its timber generally of poor quality and growing upon inaccessible mountains ... its fur trade ... of insignificant value, and will speedily come to an end ... the fisheries of doubtful value ... in a climate unfit for the habitation of civilized men."

Today, Alaska supports a population of more than one half million people and an annual economy worth more that $9 billion.

THE BULLETIN INTERVIEWS: Professor H. E. Landsberg[1]

H. Taba

Helmut Erich Landsberg was born at Frankfurt-am-Main in February 1906. At Frankfurt University he studied a wide range of Earth sciences and the subject of the dissertation he defended for his Ph.D. was seismographic recorders. Through his former tutor, Professor Beno Gutenberg, Landsberg was offered a post involving teaching and research work in geophysics and meteorology at the Pennsylvania State University, and he went to the USA in 1934. In 1941 he was appointed to the faculty of the University of Chicago which had a particularly high reputation among atmospheric scientists of the day, the Meteorological Department being headed by Professor Carl-Gustav Rossby. In 1949 he became a member of President Truman's Air Pollution Committee and contributed to the formulation of air pollution legislation.

It is not without reason that many meteorologists consider Professor Landsberg as the father of modern climatology. He wrote a book, *Physical Climatology*, published in 1941, which demonstrated clearly that climatology was not simply an exercise of geographical classification by statistics, but a well-developed applied physical science. His use of statistical analysis in climatology was accepted in the scientific community and caught on rapidly. Next, he embarked on a completely new field, namely the effect of urbanization on the environment, in particular the local climate. He was also in charge of a project which applied climatological knowledge to military strategic planning. From 1954 to 1967, Professor Landsberg served as director of the U. S. Weather Bureau's Office of Climatology, where he introduced computer methods and encouraged collaboration between the Weather Bureau and universities. He participated in the high-level discussions which laid the foundations for the establishment of the Environmental Sciences Service Administration in 1965 and subsequently the National Oceanic and Atmospheric Administration in 1970. Already in 1962 Landsberg had joined the faculty of the University of Maryland, where he organized a graduate programme in meteorology; in due course he left the Weather Bureau and became director of the university's Institute for Fluid Dynamics and Applied Mathematics. In 1976, he was appointed professor emeritus and still guides scientists and students in their work.

Professor Landsberg has distinguished himself highly in the international arena. He served on several WMO working groups and in 1969 was elected president of the Commission for Climatology and held office for the maximum period of two terms (eight years). He conducted the affairs of the Commission with utmost skill and efficiency. His farsightedness convinced the meteorological community that much higher priority needed to be given to climatology, and this led to the launching of the WMO World Climate Programme in 1979.

---

[1] *WMO Bulletin*, **33**(2), Apr. 1984

F. Baer et al. (eds.), Climate in Human Perpsective, 97–109.

Professor Landsberg is the author of more than 300 articles and monographs as well as climatological reviews and encyclopedia articles. His services to the American Geophysical Union began over 30 years ago and continue without interruption. He was chief editor of the fifteen-volume series *World Study of Climatology* and has himself written four books. Moreover, in his private library he has a collection of almost 300 volumes of early meteorological literature, especially seventeenth, eighteenth and early nineteenth century works. Readers will be interested to know that he intends to write a series of brief retrospective reviews of these books, and thereby to assess their impact on the development of the science of meteorology. Professor Landsberg is a member and fellow of several associations, academies, institutes and societies and he has received numerous recognitions and awards, including the IMO Prize. Notwithstanding his renown, perhaps the most endearing quality about Helmut Landsberg is his amiable character and willingness to help. In 1978, when he received the William Bowie Medal for outstanding contributions to fundamental geophysics and for unselfish co-operation in research, he modestly said that the citation gave him far more credit than he deserved. There is no doubt that Professor Landsberg's contributions to science and society will be remembered throughout the world and for a long period.

The Editor is most grateful to Professor Landsberg for giving him the opportunity for this interview which took place on 29 July 1981.

H.T. - **Please would you tell readers something about your childhood and family background.**
H.E.L. - I was born in 1906 at Frankfurt-am-Main in what is now the Federal Republic of Germany. My father was a physician and my mother a housewife, but with a good education. During my childhood I have happy memories of frequent visits to relatives in the countryside, and a couple of trips to Italy. But there was also the First World War with its terrible food shortages and frequent air-raids. We all suffered from malnutrition; this period was a traumatic experience for me. I had no brothers or sisters, but plenty of friends, and being a Boy Scout ensured that I had plenty of activities.

H.T. - **Which university did you go to, and what subjects did you take?**
H.E.L. - I went to the University of Frankfurt and studied mathematics, physics, meteorology, geophysics and geology, also sitting in on some courses in philosophy and biology. I obtained a Ph.D. in natural sciences. My thesis actually was in seismology; I made a comparison of two earthquake-measuring instruments. However, I took the full course in meteorology which included some practice forecasting. I stayed on at the University as post-doctoral assistant to Professor Franz Linke who had the Chair of Meteorology. One of my first jobs was to set up an observing network in the Rhineland uplands to study how to prevent frost damage to the vineyards in that region.

H. T. - **From 1930 to 1934 you were the supervisor of the Taunus Observatory. Where was this, and what work did you do?**
H.E.L. - The Taunus Observatory was part of the University's Institute of Meteorology and Geophysics. It was high up in the Taunus Hills, about 800 metres above sea-level, well away from the smoky atmosphere of Frankfurt. It was a fully-fledged meteorological

observatory; there was even a kite station although it was no longer in use. There was a seismological station too. Our main work was making observations in radiation and atmospheric electricity, and also some investigations on condensation nuclei in the atmosphere.

H.T. - **In 1934 you went to the United States of America where you were appointed Assistant Professor of Geophysics at Pennsylvania State College. What made you decide to go to the USA?**
H.E.L. - The professor who had supervised my thesis had gone there in about 1930, and I had casually said to him that if he ever heard of a vacant post I would like to go there too because I wanted to learn English. Thus it was that in 1933 I got a letter from the Dean at Penn. State saying that Professor Gutenberg had recommended me as a young and capable geophysicist and asking whether I was still interested in a position. I wrote back and said I would be glad to come, and that was that.

H.T. - **Could you please tell me something about what you did at the Penn. State? How long did you stay there?**
H.E.L. - I stayed for nearly seven years and established both the geophysical and meteorological laboratories. I was able to bring in some other faculty members to do some interdisciplinary work, and, while my initial appointment had been only for one year, it was renewed and I got an associate professorship. We were attracting quite a number of students; as a matter of fact one of them was George Cressman, who later became head of the National Weather Service.

H.T. - **From 1941 to 1943 you were Associate Professor of Meteorology at the University of Chicago. What was the state of meteorological education there in those days?**
H.E.L. - The University of Chicago had just established a department of meteorology under Professor Carl-Gustav Rossby, and it was he who invited me to join the department. My other colleagues were Horace Byers (also an associate professor), Michael Ference and Victor Starr who later went to MIT. We were then engaged in a very extensive programme to train meteorologists for the United States Army Air Corps (as it was then called -- now it is the United States Air Force). There were several other universities doing the same thing. It was a very concentrated course; my particular teaching job at that time was a field course in which we acquainted young cadets with instruments and instrument handling. With a large group we did an interesting three-dimensional study on the lake breeze at Chicago--something which was quite unusual in those days. I also gave a course in climatology because most of the cadets had never been out of the USA and had no idea what the rest of the world looked like.

H.T. - **Was this the start of your interest in climate, or had you already done some work in climatology?**
H.E.L. - My interest started back in Frankfurt because the post-doctoral work that I did was partly climatology; Professor Linke had a climatology project at the University in which I supervised a number of graduate students who were evaluating and processing data. Besides that, I did a number of other studies dealing with climate, both in Germany

and at Penn. State, although at Penn. State my interests were perhaps a little closer to problems of air pollution.

H.T. - **In those days air pollution was hardly regarded as a problem. How did you come to study this topic which has become such a burning issue nowadays?**
H.E.L. - I always had a fundamental interest in cloud physics, and of course pollutants were involved in this. I was also interested in biometeorology - the effects of weather on health--where pollutants might play a part. Perhaps the medical aspect came from my family background. I wrote a monograph on atmospheric condensation nuclei which was published, and then a book on physical climatology (at that time there was no up-to-date textbook on climatology available in the USA). It was first published by the Penn. State Press, and later came out in a much larger edition after the Second World War.

H.T. - **Now you were working closely with Rossby, whose interest was purely in atmospheric dynamics. How did he react to someone with such diversified interests as you had?**
H.E.L. - He and I got along famously because at that time Rossby's interest was in boundary-layer meteorology. He had earlier done some work with Montgomery at MIT[2] that dealt with the boundary layer and we were given some money to tackle a practical problem that involved the study of dunes on the eastern side of Lake Michigan. I did the field work, and then Rossby and I figured out the theoretical interpretation. We had a very harmonious relationship.

H.T. - **Apart from the trainees for military purposes, did you also have regular courses?**
H.E.L. - These courses were sort of mixed together. Yes, we did have civilian students, some of whom have since become quite well known like George Platzman and Joanne Gerould, better known today as Joanne Simpson.

H.T. - **This was before Palmén, Nyberg and other Scandinavians visited the University of Chicago?**
H.E.L. - Yes. The war was still in progress then of course and one could not move around easily. Nevertheless, we did have Jack Bjerknes and Sverre Petterssen, both of whom I came to know well. After the war we had lots of contacts with the Europeans and of course Rossby himself had been at MIT since the middle or early 1930s and had built up a thriving department there. I can only assume that he left MIT because the University of Chicago offered him a better position. It was a well-endowed and prestigious university with several Nobel Prize-winners on the staff. Rossby was a dynamic person, continuously on the move, always with a new scheme for something or other. He was a stimulating personality and I enjoyed my contacts with him tremendously.

[2]Massachusetts Institute of Technology

H.T. - **During the Second World War you were a special consultant and analyst for the United States Air Force in Washington, D.C. What did your duties consist of?**
H.E.L. - I started to do some consulting work for the U.S. Army Air Corps in February 1942. My main work was to inform military strategists about climatic conditions in various far-flung theatres of war so that tactical planning took this important factor into account. In fact the Air Corps sponsored a military climatology project at the University of Chicago which I supervised for them; a very large group of people were analysing data, and we also looked into possibilities of extended-range forecasting which was of interest to the military authorities. Later I started working full time as operations analyst for both the Eighth and the Twentieth Air Force and visited the scenes of action for a considerable length of time to see how the weather influenced air operations. I wrote several reports which of course were then given very restricted circulation for security reasons.

H.T. - **Which parts of the world did you visit in conjunction with this?**
H.E.L. - First I visited the air route from North America via Greenland and Iceland to Europe. After that I went to parts of North and West Africa, then across to Brazil to look at that air route, and home via Puerto Rico. The idea was to acquaint us with what was going on, and to see whether one could establish weather statistics in relation to various types of operations. I did a study also on the shuttle operation between England, Italy and the USSR.

H.T. - **What sort of data did you have to work on?**
H.E.L. - Well, the Air Force had started to put together the northern hemispheric charts prepared by the Weather Bureau. At that time there were available to us about 40 years of daily hemispheric charts which we used to compile basic statistics; we also had the *World Weather Records*, and lots of other climatological data publications. This was the time we began to put information on punch cards so that they could be machine tabulated.

H.T. - **Were these efforts in the Air Force co-ordinated with similar work elsewhere, in say the Weather Bureau or the U.S. Navy?**
H.E.L. - There was certainly co-operation and some co-ordination. This was effected through the Joint Meteorological Committee consisting of Dr. Francis Reichelderfer as Chief of the Weather Bureau, Captain Howard Orville as head of the Naval Weather Service, and Colonel Donald Yates as head of the Air Weather Service.

H.T. - **After the war, in 1946 you were appointed deputy and acting executive director of the Joint Research and Development Board's Committee on Geophysical Sciences, and subsequently executive director when it became the Committee on Geophysics and Geography in 1949. What was the function of this committee?**
H.E.L. - The Joint Research and Development Board grew out of the wartime activities. It was first headed by Dr. Vannevar Bush, a very famous scientist, who was essentially the scientific adviser to the President. Dr. Bush felt it was necessary to have not only a mechanism to co-ordinate research activities but also to do large-scale planning in various fields. I believe there were twelve different committees dealing with specific subjects, and one of these was the Committee on Geophysics and Geography whose main function

was to bring together people from the academic world, from private institutions, research institutions and from various government agencies, to lay down comprehensive plans and co-ordinate activities as far as possible. An important accomplishment of that committee was to set up what was originally called the Snow, Ice and Permafrost Laboratory, now known as the Cold Regions Laboratory. It is run by the U.S. Army in New Hampshire now, after originally having been in Chicago. Another thing we did was to try to re-establish contacts with scientists in the former enemy countries to help us out with our research and acquaint us with their own research during the war.

H.T. - **From 1951 to 1954 you were head of the Geophysics Research Directorate of the U.S. Air Force. Could you please describe what you were doing?**
H.E.L. - The Air Force Research Centre at Cambridge (Massachusetts) had two research directorates. One was for electronics and the other for geophysics. Out task in the Geophysics Research Directorate was to do research and sponsor research activities in the various fields of geophysics in universities. I had seven laboratories doing research, some of which were 'classified' (i.e. secret). I remember that we started, under the code-name 'Moby Dick', the first constant-level balloon programme, and Vincent Lally, who is now at NCAR, contributed greatly to that. We also started the first numerical prediction unit under the then Major Philip Thompson, and I charged him with running it in real time to see whether it could be done. Later the group was transferred to Washington to become the Joint Numerical Prediction Unit. But it had all started in Cambridge.

H.T. - **So you played a major role in establishing numerical weather prediction in the U.S. Weather Bureau?**
H.E.L. - It was part of my duty as director of the GRD to see that the most advanced methods were being fostered. For example, we also sponsored research at MIT under Professor Starr on the general circulation of the atmosphere. In all we provided about US $10 million each year for research outside our own laboratories. One of the people that I brought into the Geophysical Research Directorate was Dr. Robert White; he was one of my early acquisitions of first-class people.

H.T. - **Who else did you 'discover' in this way?**
H.E.L. - I had to try to find the best available people to tackle various fields. I already mentioned Vincent Lally who did much pioneering work on constant-level balloons, and Philip Thompson who worked on numerical prediction. We had Dr. Watanabe who did much of the high-level spectroscopy and things of that nature, and Dr. Peeples Peeters was head of the seismological laboratory. What we had was a sort of 'brains trust' or 'think tank' in geophysics.

H.T. - **In 1954 you joined the United States Weather Bureau as director of the Office of Climatology. What climatological activities were pursued in the Bureau in those days? How many staff members did you have?**
H.E.L. - Dr. Reichelderfer brought me in to recognize the climatological service which was in rather poor shape. As a matter of fact, a review committee had drawn up a report entitled 'Weather is the nation's business' which was submitted to the President. The Weather Bureau came in for some criticism, and in particular the committee said that

something had to be done about climatological services. The prospect appealed to me because I always liked the idea of starting something new. So first I rearranged the data-processing systems (we had a number of weather record processing centres and I consolidated these into one major centre). Actually we acquired the first computer there -- a very primitive piece of equipment by modern standards, but nevertheless a computer. Our objectives were to get the information out promptly after the end of each month, and to this end the whole system of weather record processing was changed. I had some very capable people working for me; in Washington there was a staff of 50 or 60, and when we moved to Asheville we had an even larger group. Several hundred people were eventually got together through a co-operative system with the Air Force and the Navy which had detachments there so that they could plug into the same data source. Is think that was one of our major accomplishments; the other was the creation of state climatologists (who were normally attached to the state university) in order to disseminate climatological information within the individual states of the USA. That became a very successful scheme. Another thing we did was to establish a research unit, for which we got people like J. Murray Mitchell Jr and Harold Crutcher whom I was able to send to university to acquire a higher degree (with specialization in climatology) so as to be better qualified to serve in various capacities. Gerald Barger was another who worked with me at that time, and we also had some very useful people who came from abroad, such as Pauls Putnins who I believe originally came from Estonia; he did some very interesting studies, especially on the polar regions. Then there was a young Russian lady married to an American who did a lot of translations for us. Nobody else in our group understood Russian, so it was very useful to be able to find out what research was being done in the Soviet Union.

H.T. - **What major studies were made in climatology in the USA during the eleven years that you were in charge of the Office of Climatology?**

H.E.L. - I certainly cannot claim all the credit; there were other people working in climatology who probably did more than I did. Warren Thornthwaite, for example, deserves much credit—he was the first president of the WMO Commission for Climatology. Unfortunately, he died at an early age, but he and his collaborators did a great deal, especially in the fields of evapotranspiration, agricultural climatology and boundary-layer climatology. Thus, whilst I do not wish to make too many claims for the Weather Service, we did sponsor a number of small projects, for instance a study of the climatology of tropical storms, the results of which we published in bulletins, and some work in biometeorology in collaboration with the medical faculty of the University of Pennsylvania. One of our major efforts was to organize climatological information so that it could be used in a variety of ways, quickly selecting and putting together those types of data which were needed for a given purpose. We compiled the climatological charts for the *National Atlas of the United States of America*—that was a major accomplishment because not only were they very useful for geographers and other specialists, but they conveyed to the layman the great diversity of climates. There were a number of studies on the climate of larger cities, going into details useful for engineers designing drainage systems and things of that nature. Finally, one of our constant concerns was to co-operate with the people in the academic world so that there was a continuous exchange of information and material.

H.T. - **In 1965 you were appointed director of the Environmental Data Service within the Environmental Science Services Administration. Was this simply a new title, or were your functions substantially different?**
H.E.L. - This came about through the reorganization of activities relating to the atmospheric and oceanic environments. For a time the new institution was called the Environmental Science Services Administration (ESSA), headed by Dr. White as Administrator. Five years later it became the National Oceanic and Atmospheric Administration (NOAA). The Environmental Data Service was essentially the same as the old Office of Climatology, but with a number of new branches such as the Geophysical Data Centre, the Solar Data Centre and the Oceanographic Data Centre. My task again was to build up the infrastructure of the service now known as the Environmental Data and Information Service.

H.T. - **In 1964 you agreed to be a visiting professor at the Institute for Fluid Dynamics and Applied Mathematics of the University of Maryland. Two years later you left ESSA to become research professor there. What training and research activities were you doing?**
H.E.L. - Strange as it may seem, there had never been any university activities in meteorology in the Washington area. There was a tremendous demand for it, especially from employees in various government agencies who wanted further education in meteorology. I had the idea that maybe the time was ripe for a meteorological department to be established at the University of Maryland. So I talked to various heads of departments and others in authority at the University and the upshot was that I became a visiting professor on a trial basis, giving a few seminars on various meteorological topics at the same time as we began work on drawing up a graduate programme in meteorology. We submitted a prospectus to the Dean of the Graduate School and the University Senate and it was approved in principle. We also got a fairly substantial grant from the National Science Foundation that enabled us to bring in some other people and establish a regular graduate curriculum with authority to grant master's degrees and doctorates. The title Research Professor was a particular designation in the Institute because members were obliged to spend at least half of their time on research. The National Science Foundation gave me a grant to do some research that I had always wanted to do, namely to study what happens when you urbanize an area. The opportunity arose at the end of the 1960s when a new town was being built between Washington and Baltimore called Columbia, Maryland. We instrumented the site and then watched how the climatological elements changed as the town grew. We did this over a period of seven years.

H.T. - **In 1968 you became a trustee of the University Corporation for Atmospheric Research. What exactly does this mean?**
H.E.L. - The University Corporation of Atmospheric Research is a consortium of a number of universities (originally there were eight or nine which were selected by the National Science Foundation to administer the National Center for Atmospheric Research (NCAR) in Boulder (Colorado)). Universities which have advanced programmes in atmospheric sciences are admitted to the consortium, and we joined as soon as our doctorate programme in meteorology had been approved. I guess my colleagues felt that

I had the background and experience to be on the board of trustees administering the National Science Foundation grant for NCAR.

H.T. - **What other scientific occupations have you had since 1968?**
H.E.L. - I have been on innumerable committees, it is hard to remember how many, and they deal with a wide variety of subjects. Mostly they were committees established by the National Academy of Sciences or the National Research Council or the National Academy of Engineering. I was involved in committees on such questions as disposal of waste, siting of power plants, how to detect first signs of an effect of increasing atmospheric CO_2; for a while I was president of the American Institute of Medical Climatology--right now we are interested in the effect of ionization produced by high-tension power lines on biological phenomena. I am also still working with the group which deals with the national climate programme. These are all unpaid activities of course, but I am glad to be able to share my knowledge and experience with others in the attempt to find solutions to some of the diverse problems which face the human race today.

H.T. - **Could you evoke briefly the history of international co-operation in meteorology predating both IMO and WMO, in particular the efforts of the Florentine Academy, of Dr. Jacob Jurin and the Royal Society, the Palatinate Meteorological Society, and the Intergovernmental Conference on Marine Meteorology in Brussels in 1853?**
H.E.L. - This is essentially the history of meteorology since the invention of instruments. I've always been interested in that because I think meteorology is probably the archetype of good international co-operation. Perhaps we are forced into it because we cannot impose national boundaries on the atmosphere and one needs information on a world-wide basis. It started with the Florentine Academy after the invention of the thermometer and the barometer. At first these instruments were individually made and so were quite expensive. Only the suzerains of the day could afford to buy them, and they would only do this if they were interested in the readings. The barometer, or 'weather glass' as it was called, became very popular, and came to be used for forecasting the weather at a given point, thus being marked with qualitative criteria such as 'Stormy', 'Fair' or 'Very dry'. Later, when interest turned more towards a synoptic view of pressure and temperature, the main problem was calibrating the instruments on a numerical scale. This was only achieved early in the eighteenth century when it was realized that to compare pressure readings you need a thermometer attached to the barometer. Quite a lot of painstaking work is now being undertaken to try to infer earlier readings at various places in terms of a standard scale. As for thermometers, in the early stages there were as many as 30 different temperature scales. Another problem was to ensure uniformity in the capillaries. The winter of 1709 was exceptionally cold in all of Europe, and this fact stimulated great interest in observing the weather. People thought that something unusual had happened, and in fact we still do not know why it was such a severe winter. So in 1721 the Secretary of the Royal Society, a medical doctor by the name of Jacob Jurin, wrote a paper requesting people to make meteorological observations according to a standard procedure which he went on to describe. The results were not all that had been hoped for, probably because the observers were expected to purchase their own

instruments. The real impetus to organized international co-operation came in 1780 from the Palatinate Meteorological Society of the Mannheim Academy which had been established by the Prince Elector Karl Theodor (*WMO Bulletin* **29**(4) pp. 235-238). One of the observing stations which formed part of the Palatinate network was Hohenpeissenberg in Bavaria (Federal Republic of Germany) which is still operating today in comparable conditions, having accumulated more than 200 years of unbroken records (*WMO Bulletin* **30** (2) pp. 104-106). Stations with a similar length of records are found in Prague and Rome. Data from such stations are extremely valuable because they can be used to study climate fluctuation with some precision. Alexander von Humboldt published the first northern hemisphere climatological temperature chart in 1817 which was really a sensation. Brandes made the first synoptic analyses of the Palatinate network data and clearly established the migratory nature of weather systems.

H.T. - **When did you have your first contacts with IMO or WMO?**
H.E.L. - My first contact was with the IMO in 1946 when Dr. Reichelderfer nominated me as member of the U.S. delegation that went to Toronto in the summer of 1946 to re-establish contacts after the Second World War. I was the representative on the existing Commission for Climatology. I have been associated with that ever since--in other words, I have been in one form or another in the climate group, or whatever it was called in the various successive re-labellings, up to the present time.

H.T. - **You served as president of the WMO Commission for Climatology for two terms, from 1969 to 1978. Were you in general satisfied with the Organization's activities in that sphere?**
H.E.L. - I have mixed feelings about that. Early on we had a bit of a problem to get recognized. Climate was considered a very subordinate field compared with synoptic forecasting, atmospheric research and so forth. A group of us contrived to maintain the stature of CCl by compiling some Technical Notes which were well received and I think were of considerable value. I believe they helped ensure the continuation of the climatological work in WMO, because at one time this was in question. Later of course came the new wave of concern about mankind's activities changing the climate of the globe, and this was duly accentuated by the energy crisis, so that suddenly climate had become very important. I think the World Climate Programme was a splendid initiative.

H.T. - **Do you think it has started out on the right tracks?**
H.E.L. - Yes, I think it is a very good programme. There are so many aspects to it that I do not know exactly which one is the most important; I believe the priorities will become evident as we go along. I certainly welcome the new thrusts as regards climatic impacts; I feel these studies will lead to new techniques to gauge the real importance of climate in terms of national economics--up to now that has never been sufficiently well established. I am glad to see that they are beginning to incorporate climate into economic models. I think the World Climate Programme is going to give an economic boost to extended-range forecasting which I am sure has now become far more important than day-to-day weather forecasts.

H.T. - **In the WCP, what do you feel should be the respective roles of WMO as an organization on the one hand, and of its Member countries on the other?**
H.E.L. - WMO is essentially a planning and co-ordinating body, and I think its main function is to give guidance to individual countries as to their national climatological activities. The national problem is to get people properly trained to undertake these activities and to do the research work that is appropriate to their particular conditions. I believe the fundamental problem will be that of having available in all countries the requisite trained staff, and here WMO can help with fellowships and like support.

H.T. - **You are a member of a great many associations, academies, unions, institutes and so forth. What is the significance of Sigma Xi, Sigma Pi Sigma, Sigma Gamma Epsilon and other combinations?**
H.E.L. - These are specifically American honorary societies. You are elected for your proven efforts; Sigma Xi is a scientific research society, the counterpart of the honorary society Phi Beta Kappa on the humanities side. Sigma Xi publishes the journal *American Scientist* and arranges scientific meetings and so on. Sigma Pi Sigma is an honorary group in physics and Sigma Gamma Epsilon is in the Earth sciences.

H.T. - **In 1960 you received the Department of Commerce's highest honour, the Award for Exceptional Service, for your major technical contributions to meteorology and climatology. Could you please say something about the history of this award?**
H.E.L. - Each of the government departments has one of these awards; it is rather like the medals given for outstanding performance in one of the military services. A number of awards are made each year to employees who are judged to have done a good job. There are usually three levels: gold, silver and bronze, the 'Exceptional' Service Award is a gold medal. I do not know exactly when it was established, but presumably it has been going on for some time, certainly since the end of the war. A number of meteorologists have obtained that award in the Department of Commerce, for example I know that Dave Johnson got it for his work relating to satellites. In fact there is quite a long list. Of course it does not apply just to Weather Bureau people in the Department of Commerce, but to all engaged in economic activities.

H.T. - **You received the American Meteorological Society's Award for Outstanding Achievement in Bioclimatology in 1964. What exactly is bioclimatology, and what were your personal contributions to the discipline?**
H.E.L. - The bioclimatology award is given each year by the AMS. I believe I was the second person to receive it. The work I had done was somewhat related to WMO's activities. For example I had written Technical Note No. 22 *Preparing climate data for the user* published in 1958, and Technical Note No. 123 *The assessment of human bioclimate: A limited review of physical parameters* published in 1972, as well as a number of relatively short papers. I figured that we spend a great deal of time indoors, and the relationship between the climate indoors and outdoors was not really very well known, so I did some work on that. The AMS brought out a series of little textbooks for high school and early college students to acquaint them with atmospheric problems, and I was given the task of writing one on weather and human health. It was published and

sold quite a few copies. I wrote one paper that I have never been able to live down: it was a paper on the climate in bed. I felt that since one-third of our lives is spent in bed, we ought to know whether we are really comfortable or not. People regarded this as a great joke.

H.T. **Does the climate in bed vary as a function of geographical location? What other factors are involved?**
H.E.L. - First I must make it clear that we are speaking about a bed occupied by only one person. Yes, it does. The temperature in the bed depends on the initial temperature, in other words if you put a maximum/minimum thermometer in your bed you can find how the temperature changes as you occupy it. The temperature will depend on all kinds of variables, for instance whether the windows are open of closed, what state you were in when you entered the bed, so it is not as trivial as it may sound.

H.T. - **You have been very active in the AMS and the American Geophysical Union. What sort of work were you doing with these societies?**
H.E.L. - I served on the AMS Council (that is its governing body) for a number of years and I was vice-president for one term. I served as chairman of the Board of Meteorological Abstracts and Bibliography, an activity I consider to be extremely important because people can be informed about scientific literature only if it is appropriately indexed, and nowadays the output of literature is so rapid that there must be a journal of this type. I was really happy to be associated with that work. In the American Geophysical Union I served as president of the meteorology section, and later also as president of the Union itself. I served on the AGU translation board which publishes a number of journals containing translated papers--another very important activity because it is the only way we can discover the contributions to science of those who write in languages with which we are not familiar.

H.T. - **You have a very impressive list of published work and there is no time to go into details now, but please tell readers about the book you showed me this morning.**
H.E.L. - That was a book on urban climate. As I said earlier, I was always interested in the changes to the atmospheric environment that the human race was causing, and I have been working on that particular topic for 27 years now. I mentioned the major experiment at Columbia, Maryland when we watched the changes that took place as that area was urbanized. Finally I thought it was time to document my own experiences and the work of others, especially the people who worked on the Metromex project at St. Louis. The Academic Press was kind enough to invite me to write that book.

H.T. - **What scientific work are you doing at present?**
H.E.L. - My present work is fairly diversified, but my main efforts are on the resurrection of old weather records, in the form of either instrumental records or diaries. With a couple of students I have been working on the St. Petersburg (Leningrad) records, and we are now examining an early eighteenth century diary from New London (Connecticut) and trying to interpret some data collected in Wroclaw (formerly Breslau) in Poland.

H.T. - **Do you still maintain contact with WMO activities?**
H.E.L. - Yes, I read the *WMO Bulletin* in great detail, and I am always happy to talk with people from WMO and participate in any activities that I can.

H.T. - **What were your feelings when you first heard that you had been awarded the twenty-fourth IMO Prize?**
H.E.L. - I heard about it over the telephone from Dr. George Benton[3], I had not yet received the telegram from the Secretary-General. I was dumbfounded; I simply could not believe it. For one thing, I was sure that there were other candidates who were better qualified, and for another thing, it was unheard of to select a climatologist. So I was staggered, but delighted.

H.T. - **If you were asked to give advice to a young person who wished to take up meteorology as a career, what would this be?**
H.E.L. - This is a difficult question to answer. As an academic teacher I have certain prejudices; obviously it is quite clear that people who want to study meteorology and the atmospheric sciences have to be very good in mathematics and physics, and I think they should also begin to take a look at chemistry as well, because atmospheric chemistry is becoming a very important topic now. Statistics is another subject they should not neglect. It is unfortunate that nowadays one probably cannot cram everything into a course on meteorology, because things like numerical prediction will require a good bit of background in computer science. So some things will have to be acquired at a later stage. But one thing that I would strongly advocate is that the young person spend as much time as possible out of doors; climb mountains, go to the seashore, get a feel for the atmosphere and the environment. I cannot help feeling that a great many of our modern meteorologists do not really know how the atmosphere operates: they are rarely outside, most have never made regular weather observations, and do not know the precision of the various instruments, the accuracy of radiosonde information for instance. They see satellite pictures, synoptic analyses, and things of that nature, but I have the impression that they are not really acquainted with Nature as she actually runs. Since the atmospheric sciences deal with natural phenomena, I think that you should live with them for a while. I have been in a tropical storm on land, at sea and in the air, and I can assure you that you get a much better feeling for what is going on when you experience it first hand in this way, rather than just looking at a small squiggle on a weather map.

H.T. - **Professor Landsberg that brings us to the end of the interview. Thank you very much for according the *WMO Bulletin* this opportunity, and we wish you the best of luck and good health for the future.**

---

[3]At that time Dr. Benton was a Permanent Representative of the USA with WMO.

BIBLIOGRAPHY OF PUBLISHED SCIENTIFIC WORKS BY HELMUT E. LANDSBERG

Norman L. Canfield

This bibliography of Helmut E. Landsberg is generally limited to his original contributions to scientific and technical literature. Unless published or reprinted in a scientific journal, Congressional testimony and panel, committee, and commission reports are generally omitted. Also omitted are newspaper and popular magazine articles, letters to the editor, obituaries, book reviews, and encyclopedia articles.

Landsberg's numerous encyclopedia articles began in 1947 with a contribution to the Enclclopedia Americana Annual. These annual pieces continued until 1981. Other articles appeared in Encyclopedia Brittanica, McGraw-Hill Encyclopedia of Science and Technology, and Encyclopedia of Physics. The last contribution, to the Encyclopedia of Climatology, was published in 1987 by Van Nostrand Reinhold Co.

Landsberg's editorships of major multivolume series of books included *World Survay of Climatology* (Elsevier, 15 volumes) and *Advances in Geophysics* (Academic Press, through Volume 19, 1976).

In the following citations, Landsberg was senior author of jointly authored publications except where italics indicate a co-author was senior author.

1930

a. Erdbeben, Ursache der letzen Grubenunglücke (Did Earthquakes Cause the Recent Mine Disasters?). *Die Umschau*, **34**(46), November 15, 1930, 922-924.

b. Erschütterungsaufzeichnungen mittels eines Galvanometers als Demonstrations-versuch (Vibration Records with a Galvanometer as a Demonstration Experiment). *Z. Phys. Chem. Unterlicht*, **43**(6), 268-270, with *K. Feussner*.

c. Das Taunusbeben vom 22 Januar 1930 (The Taunus Earthquate of 22 January 1930). *Gerlands Beitr. Geophys.*, **26**(2), 141-155, with *B. Gutenberg*.

d. Das Taunusbeben vom 22 January 1930 (The Taunus Earthquake of 22 January 1930). *Natur und Museum*, **60**(4), 146-151, with *B. Gutenberg*.

e. Vergleich der Augzeichnungen zweier Galitzinpendel mit verschiedener Eigenperiode (Comparison of the records of two Galitzin Pendulums with different characteristic period). *Gerlands Beitr. Geophys.*, **27**, 326-359.

F. Baer et al. (eds.), Climate in Human Perpsective, 111–136.

1931

a. Beobachtungen zur PL-Welle (Observations of the PL-Wave). *Gerlands Beitr Geophys.* **29**(1), 1931, 64-68.

b. Der Erdbebenschwarm von Gross-Gerau 1869-1871 (The Earthquake Swarm of Gross-Gerau 1869-1871). *Gerlands Beitr. Geophys.*, **34**, 367-392.

c. Das Problem deer Erdbeben Vorhersage (The Problem of Earthquake Forecasting). *Natur und Museum*, **61**(7), 293-297.

d. Das Saarbeben vom 1 April 1931 (The Saar Earthquake of April 1, 1931). *Gerlands Beitr. Geophys.*, **31**(1/3), 240-258.

1932

a. Bemerkungen zu Dispersionssuntersuchungen bei Erdbebenwellen (Remarks on Investigations of Dispersion of Earthquake Waves). *Gerlands Beitr. Geophys.*, **35**(3/4), 370-373.

b. Erkärung der Frankfurter Wetterkarte (Explanation of the Frankfurt Weather Map). *Wetterkarte d. Offentl. Wetterdienststelle Frankfurt a/M*, **26**, Beilange, 8 pp.

c. Die Reklamewetterkurven (The Advertising Weather Curves). *Wetterkarte d. Offentl. Wetterdienststelle Frankfurt a/M*, **26**(196), July 14, 1932, 4.

d. Uber einen Fall angeblicher Erdbebenvorgefühle (On a Case of Alleged Earthquake Premonitions). *Z. Geophys.*, **8**(1/2), 107-108.

e. Untersuchungen an schnellschwingenden Pendeln in Flüssigkeiten (Investigations of rapidly oscillating pendulums in liquids). *Z. Phys.*, **79**(11 & 12), 776-786, with *A. Krebs.*

f. Wetterkunde - Wetterkarten (Weather Science - Weather Maps). *Allzeit Bereit*, **10**, Verlag Gunther Wolff, Plauen i. V., 32 pp.

g. Wetterkurven (Weather Curves). *Wetterkarte d. Offentl. Wetterdienststelle Frankfurt a/M*, **26**(160), June 8, 1932, 4.

1933

a. Die Ausbreitung des Erdbebens in der Nacht vom 20. zum 21. 11. 1932 in Westdeutschland (The extent of the Earthquake in the Nights of 20 to 21

November 1932 in Western Germany). *Wetterkatte Täglicher Wetterbericht d. Offentl. Wetterdienststelle Frankfurt a/M*, **27**(11), Jan. 11, 1933, 4.

b. Beitrag zum Thema; Seismische Bodenunruhe (Contribution to the subject; Microseisms). *Z. Geophys.*, **9**(3), 1933, 156-161.

c. Das Erdbeben in Fuldagebeit vom 15 Januar 1933 (The Earthquake of January 15, 1933 in the Fulda Region). *Z. Geophys.*, **9**(4/5), 234-235.

d. Uber Eusammenhange von Tiefherdbeben mit anderen Geophysikalischen Erscheinungen (On Interrelations of Deep-Focus Earthquakes with Other Geophysical Phenomana). *Gerlands Beitr. Geophys.*, **40**(2/3), 238-243.

e. Fliegeren und Wetterkunde (Aviation and Meteorology). *Volksflug*, **1**(1), April 1933, 2-4.

f. Gewitterwolken und ihre Bedeutung für den Segelflieger (Thunderclouds and Their Significance for the Glider Pilot). *Volksflug*, 1(3/4), May 1933, 51-55.

g. Haufenwolken als Wegweiser des Segelfliegers (Cumulus Clouds as Road Markers for Glider Pilots). *Volksflug*, 1(2), April 1933, 29-31.

h. Zur Laufzeitkurve der Ph-Welle Bei Fernbeben (On the Travel Time Curve of the Pn-Wave in Distant Earthquakes). *Linke Festschrift, Synoptische Bearbeitungen Wetterdienststelle Frankfurt a/M*, 38-39.

i. zur Seismizität des Mainzer Beckens und seiner Randgebirge (On the Seismicity of the Basin of Mainz and the Surrounding Mountains). *Gerlands Beitr. Geophys.*, **38**(2), 167-171.

j. Summen der Telegraphendrähte und Seismische Bodenunruhe (Humming of Telegraph Wires and Microseisms). *Z. Phys.*, **34**(15), 604-605.

k. Das Taunus Observatorium (The Taunus Observatory). *Südwestdeutsche Runafunk Zeitung*, **9**(42), 150 ct. p. 6.

l. Uber Tektonische und Magmatische Erdbeben (On Tectonic and Magmatic Earthquakes). *Naturwissenschaften*, 21(51), 894-896.

1934

a. A Chapter on the Research of Building Vibration. *The Penn State Engineer*, **16**(2), Nov. 1934, 4-5.

b. Die Entstehung der Erdbeben (The Origin of Earthquakes). *Natur und Volk*, **64**(9), 363-368.

c. Kerngählungen in Freiluft und Zimmerluft (Nuclei Counts in Free Air and Room Air). Bioklimatol. Beiblätter, **1**(2), 49-53, with *W. Amelung*.

d. *Nomoqramm für den Scholz'schen Kernzähler* (Nomograph for the Scholz nuclei counter). Univ. Inst. f. Meteor. Geophys., Frankfurt a/M, special print, lp., 1934.

e. Observations of Condensation Nuclei in the Atmosphere. *Mon. Wea. Rev.*, **62**(12), 442-445. (Paper with same title in *Bull. Amer. Meteor. Soc.*, **16**, 64-65, 1935.)

f. Uber die Sonnenscheinverhältnisse am Südosthang des Taunus (On the Sunshine Conditions of the Southeast Slope of the Taunus Mountains). *Der Balneologe*, **1**(2), 70-78.

g. Versuche zur Frostbekämpfung in Rheingau (Attempt at Frost Protection in the Rhine Region). *Wein & Rebe*, **16**(6), 178-191, with H. Müller.

h. Zählungen von Kondensationskernen auf dem Taunusobervatorium und auf dem Nordatlantischen Ozean (Counts of Condensation Nuclei on the Taunus Observatory and the North Atlantic Ocean). *Bioklimatol. Beiblätter*, **1**(3), 125-128.

1935

a. Bemerkung zur Frage "Jahreszeit und Organisumus" (Remark on the Question "Season and Organism"). *Bioklimatol. Beiblätter* **2**(2).

b. A Chat About the School's Earthquake Station. *Mineral Industries*, **5**(4), December 1935, 2.

c. Is the "Growing Season" a Significant Climatological Element? *Bull. Amer. Meteor. Soc.*, **16**(6,7), 169-170.

d. The Problem of Earthquake Prediction. *Science*, **82**(2115), July 12, 1935, 37.

e. Sammelreferat: Werkraumluft und Gewerbe Krankheiten (Collective Review: Air in Working Rooms and Occupational Diseases). Bioklimatol. Beiblätter, **2**(1), 35-37.

f. Some Correlations Between the Occurrence of Deep- and Shallow-Focus Earthquakes. *Trans. Amer. Geophys. Union*, **16**, 91-93.

g. On the Vertical Distribution of Condensation Nuclei in the Lower Atmosphere. *Vestnik Balneol. & Klimatol. Spol. Czskoslov.*, **15**(2-6), 330-333.

1936

a. Dust - A Hazard and a Problem in Mineral Industries. *Mineral Industries*, 5(6), Feb. 1936, 4.

b. A Genetic System of Earthquake Origin. *Trans. Amer. Geophys. Union*, **17**, 89-92.

c. Measurements with a Climatological Ultra-Violet Dosimeter in Central Pennsylvania. *Trans. Amer. Geophys. Union*, **17**, 134-136.

d. Note on Earthquake Intensities on Different Floors of Houses. *Gerlands Beitr. Geophys.*, 48, 1936, 84-85.

e. *Recording of Roof Subsidence.* Amer. Inst. Mining & Metall. Engs. Tech. Publ. No. 685, llpp.

f. Remarks on the Diurnal Variation of Earthquake Occurrence with Reference to the Helena, Montana, Swarm. *Bull. Seismol. Soc. Amer.*, **26**(3), 235-237.

g. Remarks on the Temperature Distribution in Pennsylvania. *Bull. Geogr. Soc. Philadelphia*, **34**(1), 13-17. (Abstract and discussion in *Bull. Amer. Meteor. Soc.*, **16**(10), 220.)

h. Some Aspects of Settlement and Control of Strata over Mined Areas. *Proc. Coal Mining Inst. of Amer.*, **50**, 132-139.

i. Unusual Rainbow. *Bull. Amer. Meteor. Soc.*, **17**, 384-385.

1937

a. Air Mass Climatology for Central Pennsylvania. *Gerlands Beitr. Geophys.*, **51**, 278-285.

b. Determinations of Sky-Blue. *Trans. Amer. Geophys. Union*, **18**, 143-144, with H. Jobbins.

c. The Environmental Variation of Condensation Nuclei. *Bull. Amer. Meteor. Soc.*, **18**(4/5), 172.

d. "Geophysics vs. Geology" is an Educational Problem. *Oil Weekly*, **89**, June 14, 1937.

e. Intensities of Earthquake Noises. *Trans. Amer. Geophys. Union*, **18**, 118-120.

f. Origin and Occurrence of Earthquakes. *Proc. Penn. Acad. Sci.*, **11**, 88-92.

g. *An Outline of the Methods of Geophysical Prospecting*, Mineral Industries Extension Div. Mimeog., State College, PA., 61 pp.

h. Roof Falls in Mines. *Science*, **85**(2208), Apr. 23, 1937, 407, with J. B. Merritt.

i. Some Properties of the Aerosol. *Vestnik Czskoslov. Fysiat. Spolec.*, **17**(3), 57-67.

j. Streamlining Mine Ventilation. *Mining Congress J.*, **23**(8), 22-23. (Also 1937 *Yearbook on Coal Mine Mechanization*, 304-307.)

k. The Ultra-Violet Dosimeter. *Bull. Amer. Meteor. Soc.*, **18**(4/5), 161-167. (German Translation: Das U.V. Dosimeter. *Bioklimatol. Beiblätter*, **4**(2), 79-82.)

1938

a. Atmospheric Condensation Nuclei. *Ergebn. Kosmischen Phys.*, **3**, 155-252.

b. The Clover Creek Earthquake of July 15, 1938. *Bull. Seismol. Soc. Amer.*, **28**(4), 237-241.

c. Core Testing by Radioactive Methods. *Oil Weekly*, **90**(4), July 4, 1938, 26-28, with A. I. Ingham.

d. Earthquake in Pennsylvania on July 15 - One of Three in the History of the State. *Mineral Industries*, **8**(1), Oct. 1938, 1 & 4.

e. Earthquakes and Seismology in Pennsylvania. *Monthly Bull., Dept. of Intern. Affairs*, Commonwealth of Penna., **5**(2), Jan. 1938, 3-9.

f. Eine Bett-Temperatur Studie (A Study of Bed Temperature). *Bioklimatol. Beiblätter*, **5**(2), 66-68.

g. On the Frequency Distribution of Sleet-Drop Sizes. *Bull. Amer. Meteor. Soc.*, **19**, 354-355, with H. Neuberger.

h. Measurements of Sky-Blue and U.V.E. *Gerlands Beitr. Geophys.*, **52**(3/4), 1938, 270-279, with H. Jobbins.

i. The Northern Lights of September 27, 1938. *Terrest. Magnetism & Atmos. Electricity*, **43**, 486, with H. Neuberger.

j. Relations of Travel Time Curves to the Seismic Wave Velocities in the Continental and Suboceanic Lithosphere. *Trans. Amer. Geophys. Union*, **19**, 121-124, with H. Neuberger.

k. *Some Observations on Mine-Roof Action.* Amer. Inst. of Mining & Metall. Engs. Tech. Publ. No. 934, 10 pp.

l. Static Electric Properties of a New Bakelite Plastic. *Science*, **87**(2262), May 6, 1938, 419-420, with A. I. Ingham.

m. Year 1938 Marked by Unusual Exhibitions of Northern Lights. *Mineral Industries*, **8**(3), December 1938, 1 & 4.

1939

a. Bioclimatic Tests of Ultra-Violet Radiation. *Proc. Penna. Acad. Sci.*, **13**, 110-115.

b. A Convenient Heated Precipitation Gage. *Bull. Amer. Meteor. Soc.*, **20**, 383-385.

c. Measurement of Radioactivity for Stratigraphic Studies. *Trans. Amer. Geophys. Union*, **20**, 277-280, with M. R. Klepper.

d. Measurements of Ultra-Violet Radiation Sums with Photo-Sensitive Glass. *Bull. Amer. Meteor. Soc.*, **20**, 254-256, with W. Weyl.

e. *Practical Application of Stoke's Law to Various Dust Equipment.* Safe Practice Bull. No. 41, Commonwealth of Penna., Dept. of Labor and Industry, 5 pp.

f. *Problems in Dust Determinations.* Safe Practice Bull. No. 24, Commonwealth of Penna., Dept. of Labor & Industry, 6 pp.

g. *Radioactivity Tests of Rock Samples for the Correlation of Sedimentary Horizons.* Amer. Inst. of Mining & Metall. Engs., Tech. Publ. No. 1103, 9 pp., with M. R. Klepper.

h. Roof Fall Accidents. *Mechanization - The Magazine of Modern Coal*, **3**(5), May 1939, 60-64.

i. The Speed of Cyclones over the North American Continent. *Proc. Penna. Acad. Sci.*, **13**, 116-120, with C. C. Dinger.

j. An Uncommon Method for the Determination of "g". *Science*, **89**(2307), March 17, 1939, 245-246.

1940

a. *Meteorology for Private Pilots.* Mineral Industries Ext. Div. Mimeog., State College, PA, 60 pp., with H. Neuberger.

b. *A Method of Measuring Radiation of Short Wave Length by Means of a Photochemical Reaction in a Special Glass.* The Pennsylvania State College Studies No. 5, The Pennsylvania State College Bulletin, **34**(35), 30 pp.

c. Seasonal Pressure-Changes and Earthquake-Occurrence. *Trans. Amer. Geophys. Union*, **21**(Pt. II), 227-228.

d. The Use of Solar Energy for the Melting of Ice. *Bull. Amer. Meteor. Soc.*, **21**, 102-107.

e. Weather Forecasting Terms. *Bull. Amer. Meteor. Soc.*, **21**, 317-320.

1941

a. On Climatological Aids to Local Weather Forecasting. *Bull. Amer. Meteor. Soc.*, **22**, 103-105.

b. Geo-Electric Investigations over Penn's Cove. *Proc. Penna. Acad. Sci.*, **15**, 65-68, with F. Keller.

c. Geophysics at Penn State 1934 - 1941. *Mineral Industries*, **11**(2), November 1941, 1, 3, 4.

d. *A Geophysics Option in a Comprehensive Earth-Science Curriculum.* Amer. Inst. Mining & Metall. Engs., Tech. Publ. No. 1381, 3 pp.

e. Outline of the Course in Bioclimatology at the Pennsylvania State College 1940 - 1941. *Bull. Amer. Meteor. Soc.*, **22**, 13.

f. *Physical Climatology.* First Edition, Gray Printing, DuBois, PA, 283 pp.

g. *A Review of Methods to Determine Free-Silica Content of Dusts with Special Reference to X-Ray Analysis.* Safe Practice Bull. No. 85, Commonwealth of Penna. Dept. of Labor & Industry, 5 pp.

1942

a. Developments in Exploring Lithosphere and Atmosphere. *School Science & Mathematics*, **42**(2), 167-171.

b. The Influence of Snow Cover on Air Temperature. *Proc. Central Snow Conference*, **1**, 45-49, with G. P. Cressman & H. K. Saylor.

c. A New Style of Soil Thermometer. *Bull. Amer. Meteor. Soc.*, **23**, 131-132.

d. The Structure of the Wind over a Sand-Dune. *Trans. Amer. Geophys. Union*, **23**, 237-239.

1943

a. A Program in Climatology. *Trans. Amer. Geophys. Union*, **24**, 148-150.

b. *Strategic Aspects of the Climate of Norway and Adjacent Scandanavian Regions.* Univ. of Chicago, Inst. of Meteor., Military Climatology Project, Report No. 25, 75 + 30 pp.

c. Wind Influences on the Transportation of Sand over a Michigan Sand Dune. *Proc. Second Hydraulics Conference*, Bull. 27, Univ. of Iowa Studies in Engineering, 342-353, with N. A. Riley.

1944

a. *A Climatic Study of Cloudiness over Japan*. Univ. of Chicago, Inst. of Meteor., Miscell- Report No. 15, Univ. of Chicago Press, 39 + 68 pp.

b. On the Relation Between Mean Cloudiness and the Number of Clear and Cloudy Days in the United States. *Trans. Amer. Geophys. Union*, **25**, 456-457.

1945

a. Climatology. *Handbook of Meteorology*, F. Berry, E. Bollay, and N. Beers, Eds., McGraw-Hill, 928-997.

1946

a. Climate as a Natural Resource. *Sci. Monthly*, **63**, October 1946, 293-298. (Reprinted in *Managing Climatic Resources and Risks*, National Academy Press, 1981, 17-22.)

b. On Difference of Pn-Wave Velocities. *UGGI, Assoc. de Seismol., Ser. A., Trav. Sci.*, **16**, 35-42, with H. Neuberger.

c. Note on the Frequency Distribution of Geothermal Gradients. *Trans. Amer. Geophys. Union*, **27**(6), 549-551. (Discussion by H. C. Spicer & Closing Remarks by H. E. Landsberg, *Trans. Amer. Geophys. Union*, **28**(3), 493-494.

1947

a. Critique of Certain Climatological Procedures. *Bull. Amer. Meteor. Soc.*, **28**(4), 187-191.

b. Include Climate in Highway Plans. *Better Roads*, **17**(3), March 1947, 25-28.

c. Microclimatology. *Architectural Forum*, **86**(3), March 1947, 114-119.

d. Remarks on Teaching and Research in Climatology in the U.S. *Bull. Amer. Meteor. Soc.*, **28**(9), 419-421.

e. A Report on Dust Collections Made at Mount Weather and Arlington, VA, 1 October to 20 November, 1946. *Contributions of the Meteoritical Society*, **4**(1), 50-52.

f. Tornadoes in the U.S.A., *Weather*, **2**(5), 149.

g. Use of Climatological Data in Heating and Cooling Design. *Heating Piping & Air Conditioning*, September 1947, 121-125. (Also: *Trans. Amer. Soc. of Heat. & Ventil. Engs.*, **54**, 1948, 253-264.)

1948

a. Bird Migration and Pressure Patterns. *Science*, **108**(2817), December 24, 1948, 706-709.

b. Note on Deep-Focus Earthquakes, Pressure Changes, and Pole Motion. *Geofis. Pura Appl.*, **12**(5/6), 1-4.

1949

a. Climatic Trends in the Series of Temperature Observations at New Haven, CT. *Glaciers and Climate*, Geograf. Ann., No. 1-2, 125-132.

b. Climatology of the Pleistocene (Pt. 6 of Pleistocene Research). *Bull. Geol. Soc. Amer.*, **60**, 1437-1442.

c. Prelude to the Discovery of Penicillin. *Isis*, **40**(Patt 3, No. 121), August 1949, 225-227.

1950

a. Atmospheric Dust. *General Electric Tech. Eng. News*, **32**(2), November 1950, 13, 30.

b. Comfortable Living Depends on Microclimate. *Weatherwise*, **3**(1), February 1950, 7-10.

c. Microclimatic Research in Relation to Building Construction. *Weather and the Building Industry*, Building Research Advisory Board, Research Conference Report No. 1, Washington, D.C., 23-28.

d. Study the Microclimate Before You Build. *Refrigerating Engineering*, **58**(3), 257-260.

1951

a. Alaskan Research and National Defense. *Proc. Alaskan Sci. Conf.*, Natl. Res. Council Bull. No. 122, 152-158.

b. Applied Climatology. *Compendium of Meteorology*, Amer. Meteor. Soc., 976-992, with W. C. Jacobs.

c. Climatology and Its Part in Pollution. *On Atmospheric Pollution*, Meteor. Monogr., **1**(4), 7-8.

d. Origin and Physics of the Atmosphere. *Trans. New York Acad. Sci.*, Ser. II, **13**(5), 154-159.

e. Some Recent Climatic Changes in Washington, D.C. *Arch. Meteor. Geophys. Bioklim.*, Ser. B., **3**, 65-71.

f. Statistical Investigations into the Climatology of Rainfall on Oahu (T.H.). *On the Rainfall of Hawaii: a group of contributions*, Meteor. Monogr., **1**(3), 7-23.

1952

a. Activities of the Geophysics Research Division, AFCRC. *Weatherwise*, **5**(3), June 1952, 55-58.

b. Bad Weather Made Better. *Science Digest*, **31**(4), 74-76.

1953

a. Geophysical Research. *Air University Quarterly Review*, **6**(1), Spring 1953, 63-73.

b. Klimatische Meriodionalschnitte der freien Atmosphäre. *Linke's Meteorologisches Taschenbuch, Neue Ausgabe*, II. Band, Akadem. Verlagsges. Geest & Portig K.-G., Leipzig, 379-380 with 6 plates.

c. The Origin of the Atmosphere. *Scientific American*, **189**(2), August 1953, 82-86.

1954

a. Bioclimatology of Housing. *Recent Studies in Bioclimatology*, Meteor. Mongr., **2**(8), 81-98.

b. Does the Climate Change? *Weekly Weather & Crop Bull.*, **41**(49), November 29, 1954, 7-8.

c. *Geophysics and Warfare*. Research and Development Coordinating Committee on General Sciences, Office of Ass't. Sec'y. of Defense (R & D), CGS 202/1, March 1954, 68 pp.

d. Some Observations of the pH of Precipitation Elements. *Arch. Meteor. Geophys. Bioklim.*, **7**, 219-226.

e. Storm of Balaklava and the Daily Weather Forecast. *Sci. Monthly*, **79**(6), December 1954, 347-352.

1955

a. *Climatology - A Useful Branch of Meteorology.* Royal Meteor. Soc., Can. Branch, **6**(6), 9 pp. + 2 pp. bibliography.

b. The Director's Weekly Bulletin. *Proc. 8th Amer. Conference on the Administration of Res.*, N.Y. Univ. Press, 61-62.

c. Weather "Normals"and Normal Weather. *Weekly Weather & Crop Bull.*, **42**(5), January 31, 1955, 7-8.

1956

a. The Climate of Towns. *Man's Role in Changing the Face of the Earth,* W. L. Thomas, Ed., University of Chicago Press, 584-606.

b. Industrial Climatology. *Weekly Weather & Crop Bull.*, **43**(26), June 25, 1956, 7-8.

c. Nature's Air Conditioner. *American Forests*, **62**(8), August 1956, 17, 63.

1957

a. Foreword to *Industrial Operations under Extremes of Weather.* Meteor. Monogr., **2**(9), iii.

b. Review of Climatology, 1951-1955. *Meteorological Research Reviews*, Meteor. Monogr., **3**(12), 1-43.

c. United States Water Supplies from the Sky. *Weekly Weather & Crop Bull.*, **44**(47), November 25, 1957, 6-8.

1958

a. Data Processing in Geophysics. *Contributions in Geophysics: In Honor of Beno Gutenberg*, Pergamon, 210-227.

b. Interaction of Soil and Weather. *Soil Sci. Soc. Amer. Proc.*, **22**, 491-495, with M. Blanc.

c. *Physical Climatology.* Second Edition, Gray Printing Co., Dubois, PA, 446 pp. (8 printings to 1969).

d. *Preparing Climatic Data for the User.* WMO Tech. Note 22, 19 pp.

e. Trends in Climatology. *Science*, **128**, 749-758.

f. Weather: Its Effect on Plant Location. *Industrial Development and Manufactures Record,* Conway Publications, Atlanta, GA, Oct. 1958. (Also: *Weather a Factor in Plant Location,* U.S. Weather Bureau, May 1961, 1-2.)

1959

a. Note on Frequency of High Winds Over the United States. *Mon. Wea. Rev.*, **87**, 185-187, with B. Ratner.

b. Power Spectrum Analysis of Climatological Data for Woodstock College, Maryland. *Mon. Wea. Rev.*, **87**, 283-298, with J. M. Mitchell, Jr. and H. L. Crutcher.

c. The Weather in the Streets. *Landscape*, **9**(1), 26-28.

1960

a. Bioclimatic Work in the Weather Bureau. *Bull. Amer. Meteor. Soc.*, **41**(4), 184-187. (Legends on Figs. 1 & 3 should be interchanged.)

b. *The Decennial United States Census of Climate 1960 and Its Antecedents.* Key to Meteorological Records Documentation No. 6.2, U.S. Wea. Bur., 23 pp.

c. Do Tropical Storms Play a Role in the Water Balance of the Northern Hemisphere? *J. Geophys. Res.*, **65**(4), 1305-1307.

d. Facets of Climate. *The Science Teacher*, **27**(2), 6-12.

e. Note on the Recent Climatic Fluctuation in the United States. *J. Geophys. Res.*, **65**(5), 1519-1525.

f. Some Patterns of Rainfall in North-Central USA. *Arch. Meteor. Geophy. Biokl.*, Ser. B., **10**(2), 165-174.

1961

a. Climate Made to Order. *Bull. Atom. Sci.*, **17**(9), 370-374.

b. Climatology. United States National Report 1957-1960, Twelfth General Assembly IUGG, L. H. Adams, Ed., *Trans. Amer. Geophys. Union*, **41**(2), 204-209.

c. Solar Radiation at the Earth's Surface. *Solar Energy*, **5**(3) 95-98.

d. Temperature Fluctuations and Trends over the Earth. *Quart. J. Roy. Meteor. Soc.*, **87**(373), 435-436, with J. M. Mitchell, Jr.

1962

a. The Atmospheric Sciences 1961-1971, An Essay Review. *WMO Bull.*, **11**(4), 178-182.

b. Biennial Pulses in the Atmosphere. Beiträge zur Physick der Atmosphäre, **35**(3/4), 181-194. (Weickmann Memorial Volume.)

c. Biometeorlogy of Air Pollution. *Proc. Natl. Conf. on Air Pollution*, U. S. Dept. Health, Educ. and Welfare, 149-152.

d. City Air - Better or Worse? *Symposium Air over Cities*, SEC Tech. Report A62-5, Robert A. Taft Sanitary Eng. Center, 1-22.

e. Editorial on Meetings. *Bull. Amer. Meteor. Soc.*, **43**, 675.

f. Goals for Climatology. *J. Wash. Acad. of Sci.*, **52**(4), 85-91.

g. A Note on Bedroom Bioclimate. *Arch. Meteor. Geophys. Bioklim.*, Ser. B, **12**, 159-163.

1963

a. Note on "Early Developments in the Theory of the Saturated Adiabatic Process". *Bull. Amer. Meteor Soc.*, **44**(8), 511.

b. Surface Signs of the Biennial Atmospheric Pulse. *Mon. Wea. Rev.*, **91**, 549-556, with J. M. Mitchell, Jr., H. L. Crutcher and F. T. Quinlan.

c. The Unusual Weather of January 1963. *Mon. Wea. Rev.*, **91**, 307-308.

d. *World Maps of Climatology.* First Edition, Springer-Verlag, Berlin, 28 pp. + 5 maps, with H. Lippman, K. Paffen, and C. Troll.

1964

a. Controlled Climate (Outdoor and Indoor). *Medical Climatology - 1964*, **8**, Physical Medicine Library, S. Licht and H. Kametz, Eds., 663-701.

b. Early Stages of Climatology in the United States. *Bull. Amer. Meteor. Soc.*, **45**(5), 268-275.

c. A Note on the History of Thermometer Scales. *Weather*, **19**(1), 2-6.

d. Rare Halo Phenomenon. *Mon. Wea. Rev.*, **92**(7), 352.

e. Relations Between External Weather and the Microclimate of Rooms and Buildings. *A Survey of Human Biometeorology*, F. Sargent II and S. W. Tromp, Eds., WMO Tech. Note 65, 364-371.

f. Roots of Modern Climatology. *J. Wash. Acad. Sci.*, **54**(4), 130-141.

1965

a. Climate and Health. *The Environment and Man*, Travelers Research Center, Hartford, CT, 65-70.

b. Matthew Fontaine Maury, Advocate of Weather and Crop Reporting. *Weekly Weather and Crop Bull.*, **52**(1), Jan. 4, 1965, 7-8.

c. Die mittlere Wasserdampfverteilung auf der Erde (Average Water Vapor Distribution over the Globe). *Meteor. Rundsch.*, **17**(4), 102-103.

d. *World Maps of Climatology*. Second Edition, Springer-Verlag, Berlin, 28 pp. + 5 maps, with H. Lippmann, K. Paffen and C. Troll.

1966

a. Air Pollution and Urban Climate. *Biometeorology*, **2**, Proc. 3rd Int. Biometeor. Congr., Pergamon, 648-656.

b. A Climatological Analysis of the Basel Weather Manuscript (A Daily Weather Record from the Years 1399 to 1401). *ISIS, Quart. J. History Sci. Soc.*, **57**(1), 99-101, with *R. H. Frederick* and W. Lenke.

c. Comments on Paper by D. Shaw, "Sunspots and Temperatures". *J. Geophys. Res.*, **71**, 5487-5489, with J. M. Mitchell, Jr.

d. *Interdiurnal Variability of Pressure and Temperature in the Conterminous United States.* U. S. Wea. Bur. Tech. Paper 56, 53 pp.

e. *Weather and Disease*, ESSA Tech. Note 33-EDS-1, 1-6.

f. Why Indeed Coriolis? *Bull. Amer. Meteor. Soc.*, **47**(11), 887-889.

1967

a. Climate, Man and Some World Problems. *Scientia*, **102**, May-June 1967, 1-10.

b. *Some Early Philadelphia, Penna., Temperature Records.* Inst. Fluid Dyn. Appl. Math. Tech. Note BN-495, Univ. of Maryland, College Park, 17 pp.

c. Synoptic Aspects of Mortality, A Case Study. *Int. J. Biometeor.*, **11**, 323-328, with *D. M. Driscoll.*

d. Two Centuries of New England Climate. *Weatherwise*, **20**(2), Apr. 1967, 52-57.

e. World Climatic Regions in Relation to Irrigation. *Irrigation of Agricultural Lands*, R. Hagan, H. Haise, and T. Edminster, Eds., Amer. Soc. Agronomy, Madison, WI, 25-32.

1968

a. Again - The Publication Problem. *Trans. Amer. Geophys. Union*, **49**(3), 539.

b. Atmospheric Variability and Climatic Determinism. *Eclectic Climatology*, A. Court, Ed., Yearbook of the Ass'n. of Pacific Coast Geographers, **30**, Oregon St. Univ. Press, 13-43.

c. A Comment on Land Utilization with Reference to Weather Factors. *Agric. Meteor.*, **5**(2), 135-137.

d. *Preliminary Reconstruction of a Long Time Series of Climatic Data for the Eastern United States.* Inst. Fluid Dyn. Appl. Math., Tech. Note BN-571, Univ. of Maryland, College Park, 30 pp., with C. S. Yu and L. Huang.

e. Unfinished Geophysical Business. *Trans. Amer. Geophys. Union*, **49**(4), 607.

1969

a. *Biometeorological Aspects of Urban Climates.* Inst. Fluid Dyn. Appl. Math. Tech. Note BN-620, Univ. of Maryland, College Park, 13 pp.

b. Metropolitan Air Layers and Pollution. *Garden Journal*, **19**(2), 1969, 54-57. (Also: *Challenge for Survival*, P. Dansereau, Ed., Columbia Univ. Press, 1970, 131-140.)

c. *Weather and Health*. Doubleday Anchor, Garden City, NY, 148 pp.

1970

a. Climatic Consequences of Urbanization. *J. Wash. Acad. Sci.*, **60**, 82-87.

b. Climates and Urban Planning. *Urban Climates*, WMO Tech. Note 108, 364-372.

c. Man-made Climatic Changes. *Science*, **170**, 1265-1274.

d. Meteorological Observations in Urban Areas. *Meteorological Observations and Instrumentation*, S. Teweles and J. Giryatys, Eds., Meteor. Monogr., **11**(33), 91-99.

e. Micrometeorological Temperature Differentiation Through Urbanization. *Urban Climates*, WMO Tech. Note 108, 129-136.

f. Note on Rapid Pulsations of Temperature Lapse Rate. *Bound.-Layer Meteor.*, **1**, 61-63, with T. N. Maisel.

1971

a. *The Assessment of Human Bioclimate: A Review of Physical Parameters*. Inst. Fluid Dyn. Appl. Math. Tech. Note BN-690, Univ. of Maryland, College Park, 56 pp.

b. *Atmospheric Research Needed in the Field of Energy Conversion and Use*. Inst. Fluid Dyn. Appl. Math. Tech. Note BN-691, Univ. of Maryland, College Park, 29 pp., with M. F. Eilers.

c. Interaction of Man and his Atmospheric Environment. *Selected Papers on Meteorology as Related to the Human Environment*, WMO Special Environ. Report 2, 65-72.

d. *Man-made Climatic Changes* (1971 revision). Inst. Fluid Dyn. Appl. Math. Tech. Note BN-705, Univ. of Maryland, College Park, 46 pp. (Also: Proc. 1971 Symposium Phys. Dyn. Climatol., Leningrad, 1974, 262-303.)

1972

a. *The Assessment of Human Bioclimate.* WMO Tech. Note 123, 36 pp.

b. Climate of the Urban Biosphere. *Biometeor.*, **5**, Pt. II, Int. J. Biometeor. Suppl. 16, S. Tromp, W. Weicke, and J. Bouma, Eds., 71-83.

c. Human Influence on Climate. *Läkartidningen*, **69**(23), 2772-2778.

d. Micrometeorological Observations in an Area of Urban Growth. *Bound.-Layer Meteor.*, **2**, 1972, 365-370, with T. N. Maisel.

e. Noise Increase in an Urbanizing Area. *J. Wash. Acad. Sci.*, **62**(4), 329-331.

f. Periodicities in Daily Records of Global Radiation at College Park, Maryland, U.S.A. *J. Interdiscipl. Cycle Res.*, **3**(1), 91-98, with P. A. McGrath.

g. Weather "Normals" and Normal Weather. *Environ. Data Service*, Oct. 1972, 8-13.

1973

a. Developments and Trends in Climatological Research. *Lectures Presented at the IMO/WMO Centenary Conferences*, WMO Tech. Note 130, 29-40.

b. L'influence de l'homme sur l'atmosphere. *La Recherche*, **4**(31), 125-134.

c. The Meteorologically Utopian City. *Bull. Amer. Meteor. Soc.*, **54**(2), 86-89. (Also: *Maryland*, **1**(2), 12-16.)

d. Psychological Response to Tornadoes. *Science*, **180**, 546.

e. Spectral Analysis of the Rainfall in Dakar (Senegal). Inst. Fluid Dyn. Appl. Math. Tech. Note BN-776, Univ. of Maryland, College Park, 14 pp., with *H. El-Sayed.*

f. Stadtplanung. *Promet*, **4**, 7-10.

1974

a. Anthropogenic Pollution of the Atmosphere: Whereto? *Ambio*, **3**(3/4), 146-150, with L. Machta.

b. *Environmental Conditions in a Tropical Forest Region in Thailand.* Inst. Fluid Dyn. Appl. Math. Tech. Note BN-799, Univ. of Maryland, College Park, 171 pp., with O. E. Thompson, R. Kaylor, and R. T. Pinker.

c. Hearings on World Food Supply. *Bull. Amer. Meteor. Soc.*, **55**(9), 1135-1136.

d. Inadvertent Atmospheric Modification through Urbanization. *Weather and Climate Modification*, W. N. Hess, Ed., John Wiley and Sons, 726-763.

e. Is Pollution Altering the Weather and Climate? *Environmental Engineers Handbook*, 2, B. G. Liptak, Ed., Chilton Book Co., Radnor, PA, 75-81.

f. Physical and Chemical Properties of the Higher Atmosphere. *Progress in Biometeor.*, Div. A. Pt. I, S. Tromp and J. Bouma, Eds., 46-54, 587-588.

g. Special Applications of Meteorology and Climatology. *WMO Bull.*, **23**(1), 23-25.

h. The Summer of 1816 and Volcanism. *Weatherwise*, 27(2), April 1974, 63-66, with J. M. Albert.

i. The Urban Area as a Target for Meteorological Research. *Bonner Meteor. Abhdlg.*, **17**, 475-480.

1975

a. Actual and Potential Effects of Pollutants on Climate. *Int. Conf. Environ. Sensing and Assessment*, **2**, 32-1, 1-4.

b. *Atmospheric Changes in a Growing Community*. Inst. Fluid Dyn. Appl. Math. Tech. Note BN-823, Univ. of Maryland, College Park, 53 pp.

c. The City and the Weather. The Sciences, 15, October 1975, 11-14.

d. Climatological Conditions in Sakaerat Forest, Thailand. *Geograf. Annaler*, **57**, A, 3-4, 247-260, with O. E. Thompson.

e. The Definition and Determination of Climatic Changes, Fluctuations and Outlooks. *Atmospheric Quality and Climatic Changes*, R. Kopek, Ed., Univ. North Carolina Studies in Geography, No. 9, 52-64.

f. Drought, a Recurrent Element of Climate. *Drought*, WMO Special Environ. Report 5, 41-90.

g. Effective Use of Weather Data and Long-Range Forecasting. *Proc. Soil and Water Management Workshop*, Agency Int. Development, Office of Agric., 115-135.

h. Housing: Introductory Remarks. *Environ. Health Perspectives*, **10**, 223-224.

i. Man-Made Climatic Changes. *The Changing Global Environment*, S. F. Singer, Ed., D. Reidel Publ. Co., 197-234.

j. Rainwater pH Close to a Major Power Plant. *Atmos. Environ.*, **9**, 81-88, with *Ta-Yang Li.*

k. Sahel Drought: Change of Climate or Part of Climate? *Arch. Meteor. Geophys. Biokl.*, Ser. B, **23**, 193-200.

l. Selected Cases of Convective Precipitation Caused by Metropolitan Area of Washington, D.C. *J. Appl. Meteor.*, **14**, 1050-1060, with R. Harnack.

m. Some Aspects of Global Climatic Fluctuations. *Arch. Met. Geophys. Biokl.*, Ser. B, 23, 165-176, with J. Albert.

1976

a. *Association of Meteorological Pollution Potential with 500-mb Weather Types.* Inst. Phys. Sci. Technol. Tech. Note BN-840, Univ. of Maryland, College Park, 27 pp., with D. J. Greene.

b. Characteristic Temperature spectra in a Tropical Dry Evergreen Forest Area. *Arch. Meteor. Geophys. Biokl.*, Ser. B, **24**, 243-251, with R. Pinker.

c. Climate and Land Use. *Proc. Symposium Meteor. as Related to Urban and Regional Land Use Planning*, WMO, 1-7.

d. Concerning Possible Effects of Air Pollution on Climate. *Bull. Amer. Meteor. Soc.*, **57**(2), 213-215.

e. Spectral Analysis of Long Meteorological Series. *J. Interdiscipl. Cycle Res.*, **7**(3), 237-243, with R. E. Kaylor.

f. *Weather Climate and Human Settlements.* WMO Special Environ. Report 7, 45 pp.

g. Weather, Climate and You. *NOAA* (Magazine), **6**(4), October 1976, 55-58. (Also: *Weatherwise*, **39**, 248-253.)

l. Whence Global Climate: Hot or Cold? An Essay Review. *Bull. Amer. Meteor. Soc.*, **57**(4), 441-443.

1977

a. Biometeorology. Human Response to Weather. *New York Univ. Education Quart.*, **8**(3), 8-13.

b. Climate and Shelter. *EDS, Environ. Data Service*, May 1977, 7-11.

c. The Effect of Localized Man-Made Heat and Moisture Sources in Mesoscale Weather Modification. *Energy and Climate*, Nat. Acad. Sci., 96-105, with *C. J. Hosler*.

d. Planners Take Note. *Mazingira*, **1**, 40-48.

e. Role of Atmospheric Electricity in the Atmospheric Sciences. *Electrical Processes in Atmospheres: Proc. Int. Conf.*, Dietrich Steinkopff Verlag, Darmstadt, W. Germany, 799-803.

f. Statistical Analysis of Tokyo Winter Temperature Approximations, 1443-1970. *Geophys. Res. Lett.*, **4**(3), 105-107, with R. E. Kaylor.

1978

a. Bioklimatische Behaglichkeit. *Z. Angwandte Bader- und Klimaheilkunde*, **25**(4), 405-409, with A. Gabbay.

b. Overview and Recommendations. *Geophysical Predictions*, Natl. Acad. Sci., 1-15. (Also: EOS **59**(7), July 1978, 710-718.)

c. Planning for the Climatic Realities of Arid Regions. *Urban Planning for Arid Zones: American Experiences and Directions,* G. Golany, Ed., John Wiley & Sons, 23-37.

d. A Simple Method for Approximating the Annual Temperature of the Northern Hemisphere. *Geophys. Res. Lett.*, **5**(6), 505-506, with B. S. Groveman and I. M. Hakkarinen.

e. Special Applications of Meteorology and Climatology. *WMO Bull.*, **27**(2), 103-107.

f. Useful and Useless Data for Judging Air Pollution Effects on Climate. *Air Quality, Meteorology and Atmospheric Ozone*, A. F. Morris & R. C. Barras, Eds., Special Tech. Pub. 653, Amer. Soc. Testing Materials, Philadelphia, PA, 224-231.

g. Weather Forecast: Hazy. *The Sciences*, October 1978, 6-9. (Also: *Current*, 207, Nov. 1978, 37-41.)

h. Winter Season Temperature Outlooks by Objective Methods. *J. Geophys. Res.*, **83**(C7), July 20, 1978, 3601-3616, with *R. P. Harnack.*

1979

a. Atmospheric Changes in a Growing Community (The Columbia, Maryland, Experience). *Urban Ecology*, **4**, 53-81.

b. Detailed Structure of pH in Hydrometeors. *Environ. Sci. Tech.*, **13**, August 1979, 992-994, with *D. E. Anderson.*

c. The Effect of Man's Activities on Climate. *Food, Climate and Man*, M. R. and A. K. Biswas, Eds., John Wiley & Sons, 187-236.

d. Invitation to Make Meteorological Observations According to a Common Plan. Translation of article by Jacob Jurin, together with a comment entitled Early Weather Observations in America, *EDIS Environ. Data and Information Service*, **10**(4), July 1979, 21-23.

e. *Reconstruction of Northern Hemisphere Temperature: 1579-1880.* Pub. No. 79-181 of the Meteor. Program, Univ. of Maryland, College Park, 46 pp. with *B. S. Groveman.* (Also: *Data Appendices to Publication No. 79-181.* Pub. No. 79-182 of the Meteor. Program, Univ. of Maryland, College Park, 59 pp.)

f. Simulated Northern Hemisphere Temperature Departures 1579-1880. *Geophys. Res. Lett.*, **6**(10), 767-769, with *B. S. Groveman.*

g. *Some Previously Unrecorded Observations of Sunspots During the "Maunder Minimum" and Auroras Not Listed in Standard Catalogs.* Inst. Phys. Sci. Technol. Tech. Note BN-912, Univ. of Maryland, College Park, 13 pp.

1980

a. A Bicentenary of International Meteorological Observations. *WMO Bull.*, **29**(4), 235-238.

b. *Character of the Heat Island in Baltimore, Md.* Inst. Phys. Sci. Technol. Tech. Note BN-948, Univ. of Maryland, College Park, 6 pp., with D. A. Brush.

c. Meteorological Effects of Rejected Heat. *Ann. N.Y. Acad. Sci.*, **338**, 569-574.

d. Past Climates from Unexploited Written Sources. *J. Interdisciplinary History*, **10**(4), 631-642. (Also: *Climate and History*, Princeton Univ. Press, 1981, 51-62.)

e. The Role of Reference Stations in the Search for Climatic Trends. *Ann. Meteor.* (Neue Folge), Nr. 16, Deutscher Wetterdienst, Offenbach a.M., 28-29.

f. Variable Solar Emissions, the "Maunder Minimum" and Climatic Temperature Fluctuations. *Arch. Meteor. Geophys. Biokl.*, Ser. B, **28**, 181-191.

1981

a. City Climate. *World Survey of Climatology, General Climatology*, **3**, Elsevier, 299-334.

b. *Interpretation of a Manuscript Weather Record From St. Petersburg (now Leningrad), 1725-1732.* Inst. Phys. Sci. Technol. Tech. Note BN-969, Univ. of Maryland, College Park, 26 pp.

c. *Precipitation Fluctuations in Monsoon Asia During the Last 100 Years.* Dept. of Meteor. Publ. No. 81-191, Univ. of Maryland, College Park, with *I. M. Hakkarinen.*

d. *The Urban Climate.* International Geophysics Series, **28**, Academic Press, 275 pp.

e. Using Early Weather Records. *Weatherwise*, **34**, 196-199, 203.

1982

a. Climatic Aspects of Droughts. *Bull. Amer. Meteor Soc.*, **63**(6), 593-596.

b. Climatology: Now and Henceforth. *WMO Bull.*, **31**(4), 361-368.

c. Human Adaptation to Climate. *Mazingira*, **6**(4), 20-42.

d. Which Way Climate? *Proc. Int. Conf. Climate and Offshore Energy Resources*, Amer. Meteor. Soc., 167-181.

1983

a. Climates of the Past, Present, and Future of North America. *Illinois Climate: Trends, Impacts and Issues*, Illinois Dept. Energy and Natural Resources Doc. 84-05, 2-8.

b. Climatology - The Future. *Int. J. Environ. Stud.*, **20**, 159-165.

c. *The Early Wroclaw (Breslau) Temperature Observations 1710-1721.* Inst. Phys. Sci. Technol. Tech. Note BN-1003, Univ. of Maryland, College Park, 16 pp.

d. Energy Use and the Atmosphere. *Energy and the Future*, D. MacLean and P. G. Brown, Eds., Rowman and Littlefield, 91-109.

e. Meteorological Research Needs on Floods and Their Mitigation. *A Plan for Research on Floods and Their Mitigation in the United States*, Final Report to Natl. Sci. Foundation, March 1983, 23-38.

f. The Ongoing Climate Change. *Kikuo Furuta*, Tokyo, Hylife Publ. Co., 1-17.

g. Solar Variability, Weather and Climate. *Natl. Wea. Dig.*, **7**(4) February 1983, 6-10.

1984

a. The Biometeorology of Urbanization and Housing in Developing Countries. *Int. J. Biometeor.*, **28**, Supplement, 191-196.

b. Climate and Health. *Climate and Development*, A.K. Biswas Ed., Tycooly Int. Publ. Ltd., Dun Laoghaire, Co. Dublin, Ireland, 26-64.

c. *Comments on Wolf's Notes for the "Maunder Minimum" of Sunspots and Retraction of Some Earlier Claims.* Inst. Phys. Sci. Technol. Tech. Note BN-1019, Univ. of Maryland, College Park, 29 pp.

d. Early Geophysical Data, *AGU Committee on the History of Geophysics Newsletter*, No. 5, November 1984, 1-7.

e. *Fragmentary Accounts of Weather and Climate in the Americas (1000-1670 A.D.) and on Coastal Storms to 1825.* Inst. Phys. Sci. Technol. Tech. Note BN-1029, Univ. of Maryland, College Park, 61 pp., with M. K. Douglas.

f. Global Climatic Trends. *The Resourceful Earth*, J. L. Simon & H. Kahn, Eds., Basil Blackwell, Inc., New York, 272-315.

g. Overview: Climate and Energy Interactions. *Proc. DOE/Industry Workshop on the Interaction of Energy and Climate*, M. C. MacCracken, H. Moses, and J. B. Knox, Eds., U.S. Department of Energy, Washington, D.C., 17-23.

h. Preventive Climatology. *Bull. Amer. Meteor. Soc.* **65**, 1081.

i. Variability of the Precipitation Process in Time and Space. *Sampling and Analysis of Rain*, S.A. Campbell, Ed., Special Tech. Pub. 823, Amer. Soc. Testing Materials, Philadelphia, PA, 3-9.

j. *The Weather in Wroclaw (Breslau), 1692-1721.* Inst. Phys. Sci. Technol. Tech. Note BN-1015, Univ. of Maryland, College Park, 12 pp., with M. K. Douglas.

1985

a. Climatic Data. *Handbook of Applied Climatology*, D. J. Houghton, Ed. John Wiley & Sons 1369-1431, with *D. J. Houghton*, R. Jenne, F. Quinlan, and E. Wahl.

b. Historic Weather Data and Early Meteorological Observations. *Paleoclimate Analysis and Modeling*, A. D. Hecht, Ed., John Wiley & Sons, 27-70.

c. *The History of Geophysics and Meteorology - An Annotated Bibliography.* Garland Publishing, Inc., New York and London, 450 pp., with *S. G. Brush.*

d. How State Climatology Evolved in the U.S. *The State Climatologist*, **9**(2), 10-15.

e. The Value and Challenge of Climatic Predictions. *Invited Scientific Lectures Presented at the Ninth World Meteor. Congress*, WMO No. 614, 20-32.

1986

a. Problems of Design for Cities in the Tropics. *Urban Climatology and its Applications with Special Regard to Tropical Areas*, WMO No. 652, 461-472. (This publication is "dedicated to Helmut E. Landsberg by his friends".)

1989

a. Simple and Complex Oceanic Influences on Climate: A Commentary. *The Ocean in Human Affairs*, S.F. Singer, Ed., Paragon House, New York, 135-142 (Chapter 8).

b. Where Do We Stand with the Carbon Dioxide Greenhouse-Effect Problem? *Global Climate Change*, S. F. Singer, Ed., Paragon House, New York, 87-90.

INDEX

ATMOSPHERIC SCIENCES LIBRARY

<u>GPSR Compliance</u>

The European Union's (EU) General Product Safety Regulation (GPSR) is a set of rules that requires consumer products to be safe and our obligations to ensure this.

If you have any concerns about our products, you can contact us on ProductSafety@springernature.com

In case Publisher is established outside the EU, the EU authorized representative is:

Springer Nature Customer Service Center GmbH
Europaplatz 3
69115 Heidelberg, Germany

Batch number: 10607224

Printed by Printforce, the Netherlands